Contemporary Issues in Technology Education

Series Editors
P John Williams
Curtin University, Perth, Australia

Alister Jones
University of Waikato, Hamilton, New Zealand

Cathy Buntting
University of Waikato, Hamilton, New Zealand

Contemporary Issues in Technology Education - About this series
Technology education is a developing field, new issues keep arising and timely, relevant research is continually being conducted. The aim of this series is to draw on the latest research to focus on contemporary issues, create debate and push the boundaries in order to expand the field of technology education and explore new paradigms. Maybe more than any other subject, technology education has strong links with other learning areas, including the humanities and the sciences, and exploring these boundaries and the gaps between them will be a focus of this series. Much of the literature from other disciplines has applicability to technology education, and harnessing this diversity of research and ideas with a focus on technology will strengthen the field.

More information about this series at http://www.springer.com/series/13336

P John Williams • David Barlex
Editors

Contemporary Research in Technology Education

Helping Teachers Develop Research-informed Practice

Springer

Editors
P John Williams
Curtin University
Perth, Australia

David Barlex
University of Exeter Associate
Brighton, UK

ISSN 2510-0327 ISSN 2510-0335 (electronic)
Contemporary Issues in Technology Education
ISBN 978-981-10-2817-5 ISBN 978-981-10-2819-9 (eBook)
DOI 10.1007/978-981-10-2819-9

Library of Congress Control Number: 2016957349

Printed on acid-free paper

This Springer imprint is published by Springer Nature
The registered company is Springer Nature Singapore Pte Ltd.
The registered company address is: 152 Beach Road, #22-06/08 Gateway East, Singapore 189721, Singapore

Preface

The goal of this book is to bring together significant international research in technology education by focussing on contemporary PhD theses. An international search was conducted through professional associations and higher education institutions which support postgraduate technology education research, to identify doctoral researchers who completed their theses in 2013. The long list was shortened due to availability and a range of personal reasons, and some who joined the project had to later withdraw. This left a group of 11 technology educators from 8 countries. Of course this is not a representative sample of international research in technology education, but a snapshot of doctoral research from around the world that was concluded in 2013.

It is generally accepted that no one reads a thesis after it has been examined, apart from the occasional novice doctoral student. Many doctoral students publish elements of their thesis while they are undertaking their research, often at conferences, and then focus on more substantial journal publications upon completion of their thesis. Nevertheless, the applied message of the research often does not get to the design and technology teachers who could benefit from the research to inform their practice, and therein lies the intent of this book.

Each author has been asked to write a chapter based on their thesis. Each chapter has a similar structure, with the focus being on what the research means for classroom teachers. While the conceptual underpinnings of each research project are made explicit, and the methodology is summarized, the focus of each chapter is an elaboration of the findings in ways that are relevant for practitioners. Each chapter provides a reference for the thesis so a reader can pursue more detail if they would like. Each chapter is structured around the following headings:

Abstract
The Questions I Asked and Why I Think They Are Important
How I Tried to Answer the Questions
What I Found Out
How This Might Be Used to Improve Teaching and Learning

What Might Be Investigated Further
How Teachers Might Contribute to Those Investigations

The last section of each chapter provides an indication of the authors thinking about where this research might lead in the future and, thereby, possible areas in which teachers may be interested in conducting further research.

It is the editors' wish that this book will help overcome an incongruity of educational research: it is done with the intent of improving practice, but those who do the practice rarely hear about it.

The editors would be delighted to hear from any researchers who have recently completed doctoral-level research in technology education as we hope this volume is the first in a series of "Helping Teachers Develop Research-Informed Practice" volumes and of course from any teachers who use the contents of this book to develop their practice.

Perth, Australia — P John Williams
Brighton, UK — David Barlex

Contents

Chapter 1
Introduction

P John Williams and David Barlex

International doctoral level research in technology education does not consistently get shared with the profession. Some students publish and present while conducting their research, and some progress to more significant publications after completing their theses. The publication outlets for such research vary and may or may not include technology education journals and conferences. This volume, and those that are planned to follow, represent an attempt to broadcast research findings.

The nature of this book, a collection of doctoral level research from around the world which was completed in 2013, would indicate a range of disparate chapters, although they are all within the broad area of technology education. In order to provide a coherent structure to this collection of chapters concerning the findings of technology education doctoral research, they have been grouped into four areas.

Broad Approaches to Engaging Pupils with Designerly Thinking

The first three chapters – by Farhat Ara from India, Matthew Watkins from England, and Yu Jae Young from South Korea – focus on approaches to engaging students, each reflecting the cultural context within which they are working. Each of these chapters focuses on the need to engage students in order to facilitate learning: through the development of specific pedagogical approaches, the use of engaging

PJ. Williams (✉)
Curtin University, Perth, Australia
e-mail: pjohn.williams@curtin.edu.au

D. Barlex
University of Exeter Associate, Brighton, UK
e-mail: D.Barlex@exeter.ac.uk

PJ. Williams, D. Barlex (eds.), *Contemporary Research in Technology Education*, Contemporary Issues in Technology Education, DOI 10.1007/978-981-10-2819-9_1

audio-visual introductions and through an awareness of the iterative nature of good designing.

Farhat surveyed Indian students' ideas about design and designing, and the findings suggested that most students relegated design to the status of mere decoration and associated designers with gender and professional-stereotypic images. Farhat used a constructivist theoretical framework to develop a range of pedagogical principles that could be used to engage students more deeply in their design activities and to develop more informed notions of the work designers do.

Mathew's goal was to address the documented weakness of Net Generation learners to develop deep learning in the context of social aspects of sustainability. To achieve this he developed three audio-visual workshops and collected data as the students engaged in the workshops. A number of principles emerged which can be used as guidelines for the development of introductory audio-visual experiences which will enhance student engagement and deep learning.

Jae Young's focus was to investigate the process of problem solving through the design process and then to analyze the effects on improving learning achievement. A range of data were collected from students over a period of 5 months to indicate that students followed a non-sequential iterative process, brainstorming was used frequently throughout the process and both higher and lower domains of design processes were used. He concluded that teachers need to be aware of the elements of good design so they can guide students into effective practices.

Focused Teaching and Learning in Technology Education

The second group of three chapters focuses on teaching and learning in specific areas of technology: CAD/CAM by Deborah Winn in England, Electronics by Sarah Pule in Malta and sectional drawing by Sammy Khoza in South Africa. While these are a diverse range of context areas of technology, the focus in these chapters is how to successfully teach difficult concepts. The suggestions that arise from the research include: know your students, be well informed and well planned, and teach from the real rather than from the abstract.

Deborah addresses the problem of the difficulty of mastering the teaching and learning of CAD/CAM in a context where students find it complex and teachers lack confidence when they teach it infrequently. The resource created through this research removes the need for teachers to become experts, allows students to work at their own pace and rewards effort and creativity as well as outcome. This is achieved through structural organization such as clearly defined tasks and rules, independent support for students and high expectations.

Sarah's research addresses difficulties in the transfer of knowledge in the context of electronics considering the range of student learning styles. She found that by developing specific pedagogies based on students conceptual frameworks designed to appeal to particular types of learners, their performance improved. This confirms that students' experiential work together is effective in contributing to understanding.

The problem that Sammy addressed was the difficulty of teaching and learning sectional drawing. This specific skill is indicative of the types of spatial skills students need in a range of technical professions. The model that Sammy developed from the data indicated the pivotal role of good teacher subject matter knowledge (SMK), the need for real rather than abstract models and the importance of continuous formative assessment.

Classroom Talk

The next two chapters – by Wendy Fox-Turnball and Deborah Stevens, both in New Zealand – focus on the importance of conversation, in a technology classroom with primary students and in a bioethics classroom with senior secondary students.

The data Wendy collected through conversation transcripts, student autophotographs and observations of behaviour in primary technology classrooms indicates that conversation is an important characteristic of learning for both the teacher and the students. Classroom talk enables students to deploy knowledge they have from outside the classroom and provides a conduit to use and share that knowledge. It is also the opportunity for teachers to develop deeper understandings about their students. The importance of student engagement in authentic contexts was highlighted as a contributor to motivation.

While in a different context and at a different class level, Deborah's research also indicated the criticality of student-centred narrative and discussion-based pedagogy to the achievement of skills and understandings. Surveys and interviews were used to collect data, which indicated that the narrative-based approach was effective in the development of values and key competencies and the gaining and retention of technical, scientific and philosophical knowledge in a context where there was no formal assessment and it was not required that students take notes.

Communities of Practice

The final three chapters focus on the characteristics of communities of technology teachers: Denise MacGregor in Australia, Tome Mapotse in South Africa and Gary O'Sullivan in New Zealand. These research projects consistently indicate that conceptualizing groups as communities of practice enhances their success in the achievement of their goals. The interaction of members of a community who share common concerns and goals and engage in collective learning in order to achieve those goals is a feature of each of these chapters.

Denise's research recognizes the significant transition individuals must negotiate when they move from a university community for their education, to a school community when they begin their teaching career. The data included the narratives of first year teachers' responses to a questionnaire, interviews and a reflective journal and indicated that mentoring was a key form of support when integrating into the school community. The mentoring included personal and emotional support and teaching-related assistance and advice, including feedback on professional practice. The provision of pathways into a professional learning community for new teachers could involve the use of social media, providing reduced teaching contact, establishing formal partnerships with professional teaching associations and providing beginning teachers with shared teaching and learning spaces.

Tome's research was in a South African context where many technology teachers have not been trained to teach in this area. He used action research as a way of developing competencies within a community of practice which consisted of 18 teachers. Data were collected through observations, interviews and a questionnaire; a thematic analysis of which indicated the main needs of this community were qualifications, lesson planning and assessment support, resources, curriculum interpretation and implementation, and pedagogies for high teacher-learner ratios (1:60–1:90). The action research cycle led to the development of a model (PEAR) for use in communities of practice.

The goal of the community which was the focus of Gary's research in New Zealand was to bring together school and community groups to provide education for enterprise experiences for technology education students. Data were collected through teacher workshops that were designed to develop a supportive enterprise for education community within technology education. The data, collected through observations, interviews, questionnaires and journals, indicated that student activities must be based on authentic local needs and the students must own the project in which they are engaged. The development of the community resulted in growth in teachers' professional understanding.

Chapter 2
Ideas About Design: Towards Appropriate Pedagogy for Teaching Design at the School Level

Farhat Ara

Abstract This study set out to explore Indian school students' naïve ideas about design and designers. Findings suggested that most students relegated 'design' to the status of mere decoration and associated 'designers' with gender and professional-stereotypic images. Insights from the survey afforded development of appropriate design activities and scope to extend students' understanding of design. The study has implications for development of the design curriculum at school level in India.

The Questions I Asked and Why I Think They Are Important

The teaching and learning of design is challenging when taught at the school level (Atkinson 2009). Teachers struggle to help students acquire the knowledge and skills to design as well as have a better understanding of design. This is surprising since design is something that we all engage in our daily lives from deciding what to cook for an occasion to arranging furniture for creating spaces. Design is something that has helped our ancestors to thrive and survive in severe living conditions. Humans have evolved to possess this unique ability which has enabled them to not only envisage their future but also employ their physical and intellectual resources to achieve that future (Bronowski 1973). This ability is possessed by all of us to some degree. However, what differentiates and makes school or professional design challenging from everyday design is that in case of the latter, 'design' is hidden in the act of doing and making, while in school or professional design, students/designers need to dissociate design from making.

In India design education is not organised for all. It is taught in specialised institutes providing professional design education to students only after high school (12th grade). However, in the last decade, there has been a growing concern among

F. Ara (✉)
CERTAD (Centre for Education, Research, Training and Development), Srishti Institute of Art, Design and Technology, Bangalore, India
e-mail: farhathamid@gmail.com; farhat@srishti.ac.in

PJ. Williams, D. Barlex (eds.), *Contemporary Research in Technology Education*, Contemporary Issues in Technology Education, DOI 10.1007/978-981-10-2819-9_2

Indian educators to include design in general schools (Ara et al. 2009, 2011, 2013; Choksi et al. 2006; Khunyakari et al. 2009; Mehrotra et al. 2007). My thesis emerges from these concerns around design education for all. I was curious to learn how Indian students and teachers, who had no 'schooled' understanding of design, conceptualised design. It was equally important from the curricular point of view to understand and document these ideas.

Acknowledging students' intuitive concepts has been considered significant for teaching and learning. In science education, students' intuitive concepts are well documented and have been used extensively in curriculum development (Layton 1994). There have also been similar initiatives across the globe into understanding students' ideas about and attitudes towards technology. However, not many studies have been devoted to students' ideas about design/designers. In the absence of any formal design and technology (D&T) education in Indian schools, this study aimed to contribute to broader understanding by exploring Indian students' spontaneous ideas of design so that appropriate activities can be developed for school students in working towards an experiential understanding of design.

The broad aims of my thesis were thus to investigate Indian school students', teachers' and designers' ideas of design, develop design-based activities through trials among middle school students and assess the influence of these activities on their ideas of design. In this chapter, I intend to focus on the investigation of students' ideas of design and discuss the development of design activities based on the investigations. The following questions guided my inquiry into students' perceptions and designing of the activities:

- What are Indian elementary and middle school students' ideas and images of design and designers?
- What activities, products, professions, skills and qualities do they associate with design and designers?
- What specific activities can be developed for Indian middle school students to engage them in design?

How I Tried to Answer the Questions

In order to explore and gain insights into students' ideas and images of design, I used a survey research method which enables mixed methods of data collection and analysis. For example, questionnaires with descriptive, pictorial and drawing tasks, and open-ended and closed-ended questions, as well as the interviews, ensured participants' expressions of their ideas through a range of means.

The Survey

The survey involved selecting the student sample, developing the questionnaire, administering it and analysing the responses. The questionnaire was checked for clarity and appropriateness of language and pictures in terms of age, gender, context and logical content validity. The survey was conducted with 511 students from classes 5 to 9 at an urban school. In addition, a sample of 22 students (classes 7–9) from a similar school was interviewed. Both the schools were coeducational, consisting of almost equal number of boys and girls. Students came from different linguistic backgrounds.

The Tools

A review of literature discovered very few studies aimed at understanding students' conceptions of design; and most were limited to students who had an exposure to the process of design. Two studies served to inform the construction of my questionnaire. The first was by Newstetter and McCracken (2001) with computer science and engineering students and the other with Canadian school students by Welch et al. (2006). Based on this literature, I designed a scheme that guided the development of two sets of questionnaires, one each for the elementary and middle school students. The questionnaires included:

Section A: A short description to the purpose of the survey and collected information on students' name, age, gender, etc.
Section B: Required them to 'draw a designer' at work; they were also asked to mention the gender, location and activity of their designer.
Section C: Included questions pertaining to aspects of design and designer: open-ended, sentence completion and dichotomous questions.
Section D: Consisted of several pictures showing activities performed by individuals; students had to mark those which they considered were designing activities.

To maintain gender parity, I developed two versions of Section D – one in which all activities shown were done by males while the other depicted all activities done by females. Each student randomly received only one version of this section. Section C was made simpler and shorter for younger students. I conducted the interview sessions with middle school students from classes 7 to 9 in a different school using the same questionnaire. These interviews were aimed at detailed exploration of their ideas and were audio recorded.

What I Found Out

For the sake of brevity, I have consolidated the analysis based on students' written and interviewed responses to the open-ended and close-ended questions and the drawing task. I will refer to the differences in students' responses due to age/gender only when they are significant.

Responses to the Open-Ended Questions

The analysis of these responses revealed that although these students did not have formal learning in design, they had prior ideas about what design is and what designers do. Most of them associated design with art, mostly painting, decoration and drawing/pattern making. Other ideas infrequently associated with design were making things, planning, inventing and ideating. Students also suggested examples of designed products and design professionals mostly from two design domains – fashion/dress designing and architecture. Elementary school students held simpler ideas, mostly design as painting/drawing, compared to the older students who had a more varied understanding of design, namely, design as making, planning, ideating or shapes of things.

Probing in the interviews suggested that students considered design as a vehicle for self-expression. Even when students mentioned drawings, they considered these as a tool used by designers to communicate with themselves than with others such as makers or clients. According to Butler (2012) art is a vehicle for self-expression as it first serves the artist and then serves the others just by existing or by inspiring. Design, however, never first serves the designer; it first serves the client and the process of serving the client serves the designer. That a designer works under constraints and for specific users was almost absent from the responses of all students. Owen (2005) suggests that the perceived similarity between a designer and an artist stems from their common use of the visual media to communicate ideas. However, their fundamental methods, results and goals are different.

Only a few older students made reference to two steps of the designing process (i.e. planning/ideating and making) in their spontaneous responses. When asked to suggest words/meanings for the word 'design' in different Indian languages, students suggested words strongly associated with art.

Most older students agreed that animals designed while younger students considered the contrary. Those who agreed mostly provided examples of the home-building activities of birds. Interestingly while considering design in general, students mainly thought of design as some artistic rendering process, while in the case of animals, they emphasised on their 'making' skills. Younger students ascribed designing abilities to animals on the basis of their ability to make marks on the ground with their feet/paws indicating their strong association with pattern making or decoration.

Responses to the Structured Questions

Structured questions included probing questions on the nature of design, students' attitude and interest towards design/design learning, cocurricular school subjects/activities related to design, suitability of a design profession to boys/girls and skills and qualities of a designer.

(a) *Design and school subject/activities*

When asked whether given cocurricular subjects/activities (pictorial activity in Section D of the questionnaire) were associated with design, students considered those subjects/activities which involved a more 'design as planning' approach and the design of intangibles as non-designerly. For example, subjects such as library management, music, drama and gardening were considered non-designerly. Similarly, activities such as arranging books on a shelf, teaching and cooking were considered less designerly by students. None of the students considered that each activity could be turned into a design problem if it is posed within a problem-solving context.

(b) *Design and stereotypes*

When asked whether a particular profession was more suitable for a girl, a boy or both, students considered professions such as cooking, teaching, jewellery designing, fashion designing, tailoring and interior designing as more suitable for girls as opposed to carpentry and mechanical and automobile engineering, which were considered more suitable for boys.

A list of ten statements was used to probe students' interests and attitudes towards design. Overall students showed a positive attitude towards designers and design learning. In another set of questions on qualities of a designer, they most frequently chose positive qualities such as 'intelligent', 'hard-working', 'friendly', 'kind', etc. This finding resonates with Welch et al.'s (2006) study where students had a positive attitude towards design and designers. However, unlike the findings in Welch et al.'s sample who suggested that designers could be either male or females, students in my sample portrayed a gender stereotype by choosing 'females' more frequently as designers than 'males' from among the two options. They also considered that girls were better designers than boys. Interestingly gender stereotypes of this kind were more frequent among girls than boys. More girls also held a positive attitude towards design learning. However, it appears that their positive attitude was aligned more with their idea of design as an artistic process than as problem-solving.

Students' Responses to 'Draw a Designer at Work'

In their drawings, students primarily conceptualised a designer as a fashion designer, artist, architect, engineer and a few as a labourer and a scientist. Students' understanding of design as art also gets expressed in their drawings where students (mostly younger) depicted artists engaged in some artistic work. Students' descriptions of gender and professional stereotypes of designers are evident more in their drawings than in their written responses since many older students depicted female fashion/dress designers (Figs. 2.1 and 2.2). Interestingly, these stereotypes seem to develop with age since very few younger students depicted female dress designers.

Younger students, however, seemed to conflate artists with designers. The products of design in students' drawings were mostly dresses, buildings or art works.

Fig. 2.1 A female designer dressing up a female model (class 8, girl)

Research studies on students' attitudes towards technology have found students' strong association of technology with computers and electric/electronic equipment (Khunyakari et al. 2009; Jarvis and Rennie 1998; Mehrotra et al. 2007). Even students' drawings of engineers revealed depictions of machines, vehicles, rockets and robots (Knight and Cunningham 2004; Fralick et al. 2009). In this study however, students did not depict/mention any electrical/electronic equipment as designed products. This observation can be extended to the analysis that students were not able to see any strong link between design and technology. Responses in the interviews suggested that any link between design and technology was superficial since they generally considered design as providing aesthetics to technological products.

A large number of students' depiction of dress/fashion designers may be the influence of the association and use of the word 'design' with dresses. Colloquially, the word design is used to represent patterns of or on dresses. It could also have been due to the prevalence of only a few kinds of designers in the popular media, such as fashion/interior designers. No wonder adolescents form a strong link between design and fashion. Even Owen (2005) suggests that there is confusion among the general public about the nature of design due to the extensive use of the word 'design' to mean fashion. While fashion in attire is judged predominantly on appearance, functionality and performance do play a part in fashion design. The role of functionality and performance have a more dominant role to play in other design professions where aesthetics have to some extent a lesser but still significant role.

The analysis of designerly activities depicted by students suggested 'sketching' as the predominant activity (Figs. 2.2 and 2.3). Very few students mentioned the work of designers as planning or modelling. An understanding of the role of modelling and testing as a part of design practice is fundamental to design thinking. Although they attempted to depict some form of testing or evaluation of design in

Fig. 2.2 A young female designer 'painting' a dress (class 9, girl)

Fig. 2.3 A young male designer, painting in a village (class 9, boy)

their drawings, these were all strikingly related to fashion designing (Fig. 2.1). The fact that designers work with clients and customers was depicted in some drawings (Fig. 2.1). However, that a designer works with the 'maker' was depicted in only a few of the drawings. Designing was mostly viewed as a solitary act since none of the designers were shown to work in teams (Figs. 2.2 and 2.3). The designers were shown dressed neatly and often trendily (Fig. 2.1). The profession of design seemed to be associated with a lucrative, hobby-like soft skilled profession that students assumed are pursued mostly by females. It is important to note that while the activities of science/engineering are usually perceived to be restricted to laboratories (Mead and Metraux 1957; Chambers 1983) or classrooms (Picker and Berry 2000), designing seemed to be perceived as everyday acts done at home, as well as 'white-collar' jobs done seated in offices. Designing, as students understood, was also reflected in the work of people around them such as a homemaker, a tailor, an artist (Fig. 2.3) and even labourers. This perception might be fruitful since students considered designing as something which is accessible and done by all. However, it restricts their perception about design to something which is trivial and superficial and done by females.

How This Might Be Used to Improve Teaching and Learning

In the next phase of my thesis, I drew upon the findings of the survey on students' ideas and developed strategies for teaching design to middle school students. My effort in this section will be to briefly describe my theoretical framework and then highlight the pedagogical principles that I derived from the framework in a manner which would support teaching and learning design to students who have had no formal design education.

What can be the best way to teach and learn design? It seems obvious that the best way to learn design is to 'do design'. This approach is also characterised as the experiential learning approach or learning by doing, which underlies the assumption that experiences gained through working get translated into meaningful insights into the subject matter, which in turn leads to increased curiosity and motivation for further learning among students. However 'doing design' presupposes a theoretical framework which guides an educator to take possible courses of actions which would engage students in design activities.

Theoretical Framework for Developing the Design Activities

A theoretical framework serves as a base for the study and provides rationales for the research methods used. The activities that were developed drew upon several contemporary theories of learning design in schools, particularly Robert's (2005) four roles of design in the classroom and Barlex and Rutland's (2004) pedagogical approach to design as a decision-making process.

Roberts suggested that in the classroom context, design can be viewed through four different but complementary perspectives: the designer, the maker, the user and the observer (Fig. 2.4). These roles provide working perspectives towards better comprehension of design and technology activity and of cognitive modelling. As observer and user, students are involved in judgement and evaluation of existing realities, while as designer and maker, students plan, model, test, evaluate and make artefacts.

The roles of designer and maker involve overt modes of design activity wherein students plan, make mock-ups, test, evaluate and make artefacts. The roles of user

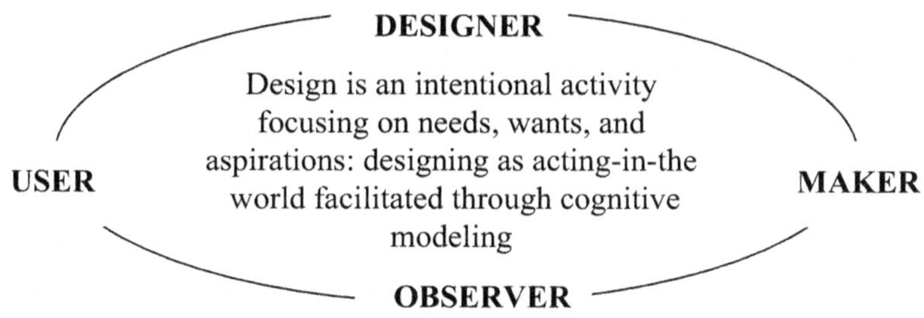

Fig. 2.4 Four roles of students in a design classroom allowing complementary perspectives on learning through designing (Roberts 2005)

and observer, on the other hand, involve covert modes of design activity wherein students get involved in continuous judgement and evaluation of existing realities and state of affairs. These two modes are complementary to each other. Roberts asserts that the covert form of design activity is often undervalued in schools. He argues that this activity provides an understanding of design in its entirety and is equally fundamental in professional designing.

Barlex and Rutland (2004) in their pedagogical approach to design as a decision-making process proposed that students need to make at least five different types of design decisions while engaging in a design task. These interconnected decisions are *conceptual decisions* which require students to think about the overall purpose of the design, *technical decisions* which require students to consider 'how the product will work' and the nature of the components and materials required to achieve this, *aesthetic decisions* which involve students to think of 'ways in which the product will appeal to the senses', *constructional decisions* which require the student to consider 'how the product will be made' and the tools and processes needed to achieve this and *marketing decisions* which require the students to consider 'who the product is for', 'what is its cost', etc. Apart from these, other theories that informed my framework were Piaget's constructivist theory of learning, Papert's theory of constructionism and Vygotsky's sociocultural theory of learning. The following subsections highlight the key pedagogical principles or approaches that I derived from the theoretical framework.

Pedagogical Principles

Widen Not Negate Students' Ideas

In a constructivist classroom, a teacher begins where the students are and then builds a conceptual bridge between their prior knowledge and where the students need to be. Students are not passive receivers but actively construct their own knowledge based on their prior experiences in the face of new experiences or information. Thus it is worthwhile for a teacher to value these initial ideas. Students came with varied ideas about design and designers in my classroom. For example, as suggested in the previous section, students more strongly associated design with art practices/products than with technology or creative problem-solving processes. They also seemed to disregard the purpose and constraints of design. Even if students considered the purposes for designing, frequently those were related to employing aesthetic appeal. Most students associated design with female dress designers and a hobby-like activity performed in isolation. These were not considered as incorrect ideas, and the main purpose of designing the activities was not to negate any aspect of the range of ideas that students understood by the term 'design' but to extend and broaden this range. The intention therefore in any such study should be to refine students' ideas, not replace them. Why I chose the four roles model was because it provided students opportunities to understand design through multiple viewpoints. By adopting these roles in the classroom, students engaged in different aspects of the design process, thus getting a wider perspective of what design is and what designers do.

Design Diverse Experiences Ranging from Familiar to Unfamiliar and Certain to Uncertain Contexts

Beginning with familiar contexts provides students possibilities to question their own assumptions and clarify their understanding in a nonthreatening way. It allows them to feel safe and confident in taking risks within known circumstances. One of the contexts that teachers can make use of while introducing design is the use of products that students are familiar with. While building the framework, working with familiar products would also allow students to assume the roles of observers and users and help them understand the designed world, an important outcome for its own sake. It also enables them to develop skills and knowledge that assists them in their designing and making.

In my model which I adapted from the Roberts model, I integrated a level of familiarity and certainty in the activities. The level of certainty can be considered to be high when one is familiar with the situation in hand. Designing is a multidimensional activity, and one of the inevitable ingredients of a design task is uncertainty, which designers need to manage through rigorous planning and decision making (Kimbell and Perry 2001). The tasks thus can progress from increasing to decreasing levels of familiarity and certainty.

Keeping the levels of familiarity and certainty in mind, I identified diverse design activities that students would engage in to assume each of the roles in the Robert's model (Fig. 2.5). For example, the role of 'user' was identified and related to the activities of handling familiar and unfamiliar artefacts and reflecting on the history of a familiar artefact. The roles of 'designer' and 'maker' were related to the activities of designing a solution for a real world problem and implementing the solution through modelling, respectively. The roles of 'observer' and 'designer' were identified with the activity of actually coming up with real world problems that could be

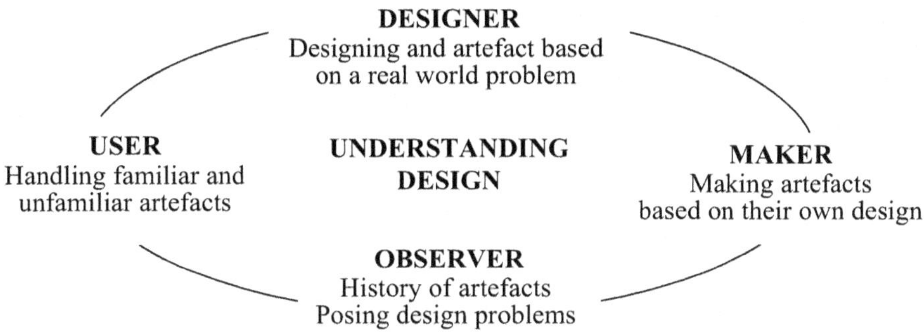

Fig. 2.5 Four roles of students in the design classroom matched with the identified design activities (Adapted from Roberts 2005)

resolved by creating artefacts. Thus the themes and contents of the tasks were formulated to meet the learning goals.

The design task sequence consisted of the following six kinds of activities in my study:

- Handling and analysing familiar artefacts
- Reviewing the history of familiar artefacts
- Handling unfamiliar artefacts, exploring and identifying them
- Designing an artefact based on a given real-world design problem
- Making artefacts
- Problem posing or looking for real-world design problems

Provide Opportunities to Make Design Decisions

Design is an iterative process where designers are required to make a series of interconnected decisions regarding every feature of the design. These decisions are not accidental but designers are conscious of their decisions needing justifications. Similarly in a design classroom, students should have abundant opportunities to make informed decisions both individually and collectively. However, students cannot make decisions in a vacuum. Hence a teacher should provide both the tools and the contexts for students so that they engage effectively in decision-making processes and also evaluate their own progress. These decisions could vary from simple ones such as the choice of colour which a student could make based on their emotional preference or complex ones such as the choice of materials, cost or durability. The key point is that students should progressively learn to be aware of their decisions and should be able to justify their choices. In my study, Barlex and Rutland's (2004) model served as a valuable tool for the students who used it for decision making, not only in the 'design and make' activities but all design-based activities such as product identification and evaluation, redesign and product sorting. As for an example, while evaluating a familiar product as a 'user', this model enabled students to understand the key decisions taken by a designer while designing that product. Again the same design decision model could be used by students while engaging in a redesigning-a-familiar-product task, where they may assume the role of a designer and take decision about the redesigned product. It thus enabled the students to make design decisions, either as users, observers, designers or makers.

Begin with Easily Available Resources in the Classroom

The design activities cannot be designed independent of resources that need to be used in the classroom. The resources available affect the design activities. One of the ways to begin design activities is by using materials that are easily accessible to the students and the teacher, during their designing and making phases. The use of everyday materials for the design tasks would also help students perceive 'design' in everyday products. For example, the everyday materials that were easily accessible and made use in my study were fountain and ballpoint pens, electric iron, pairs of tongs, etc. The other artefacts such as knife sharpeners and most of the unfamiliar artefacts were available to me in the laboratory. Besides the materials, tools and resources listed by students in their designing and making activities were provided to the students.

Set Authentic Tasks

Hennessy and McCormick (2002) emphasised authenticity of activities at two levels. At the first level is personal authenticity wherein a student has to be involved, and the learning has to be meaningful to the student. At the second level is cultural authenticity where the design task relates to the world outside of school. Providing authentic contexts to an activity will provide avenues for designerly behaviours among students while they are engaged in these activities. It has been observed that while designers work on real problems in a highly contextualised situation for which they have considerable knowledge (Anning et al. 1996) and thereby have predefined goals to resolve, the design problems in schools are artificially constructed and may not be relevant to students, thereby losing personal authenticity and becoming meaningless. Thus it is essential that teachers involve students in experiences which are not just real world but also ones which are relevant to the lives of students. This would allow students to engage in dialogue, work with real data, take action and reflect on possible outcomes, thus making personal meaning of the context.

Promote Collaboration

Working in a small group of 3–4 members in a problem-solving context provides students opportunities to work collaboratively with each other towards a shared common goal. The theoretical basis for using small groups as the pedagogic approach stems from sociocultural theories which recognise that learning is not just an individual matter, but that it develops within a social context, through interaction with peers and adults. Collaborative group work provides students possibilities to work and think together on the same problem, raise questions against each other, engage in dialogue and share their learning and insights within the secured environment of a small group of peers before sharing their work with the whole class.

In the beginning students can form their own groups based on their preference. It ensures some sort of security and comfort to begin with. A group should stick together until the completion of a set of tasks. However, as the class progresses, a teacher can increase the diversity of groups by disrupting previous group formation and form new groups based on students' generated criteria or the teacher's own criteria. It may also be favourable for students to sometimes work in single-sex groups. This prevents possible dominative behaviour on the part of boys. In my study, all the group formation was voluntary in all the trials and students naturally formed single-sex groups. One of the practices to ensure individual participation could be to ask each student to write about their contributions towards the activities, even if they are minimal. Effort should be taken to see that each student gets an opportunity to participate and share their ideas within the group; then all sorts of competitive behaviours are discouraged.

Make the Activities Gender-Neutral

Making the activities gender-neutral involves designing the tasks in a way which would be equally engaging to both boys and girls throughout the extended period of the activities. Students in my study demonstrated gender and professional stereotypes in their drawings and writings. A number of studies have also indicated that girls and boys even in the UK prefer different subjects to study and have different attitudes towards these. For example, many boys opt for electronics and resistant materials which are held in high esteem compared to the food and textiles which are mostly preferred by girls (Owen-Jackson 2013). A teacher therefore should be careful while designing activities for boys and girls. One way is to make use of a range of materials and problem-solving contexts which would help the teacher alleviate the problem associated with tasks which might have a gender bias. A teacher herself might have biased expectations about the performance of boys and girls in an activity; she therefore must be mindful and careful about her expectations.

Encourage Reflection

Although reflection is embedded throughout designing and making, fostering reflection is a key challenge to educators in D&T education. It is argued that the adherence to the design process model hinders reflective thinking by students (Mawson 2007). Reflection therefore should be consciously planned and built into design activities. Providing students opportunities to communicate their ideas to others and evaluate their own and peer ideas encourages reflection and assessment of students' own progression through the activities. Many researchers have suggested other ways of developing reflective thinking in design, for example, Lewis et al. (1998) suggest that a shift in emphasis from the design process or problem-solving approach to the problem-posing approach by students can promote reflective thinking among students. Some teachers have reported that recording of events by students in their process diary and logs can facilitate reflection (Rogers and Clare 1994), while others (Jones and Carr 1994) have suggested that an explicit instruction to reflect on their own processes and the link between the different stages can promote reflective thinking.

What Might Be Investigated Further

This study was conducted with students from urban schools in India, who did not have a schooled understanding of design or technology. Schools worldwide have technology education in their curriculum. Many of these schools even have access to computer-aided design and manufacturing and modern fabrication technology for designing. It would be worthwhile and interesting to explore what ideas and images of design and designers students with formal education in D&T hold. I must argue why I consider this question important. Students' perceptions of professions are closely related to the choice of their careers (Knight and Cunningham 2004) and images of those occupations (Gottfredson in Glick et al. 1995). If students believe that designers are artists or fashion designers and designing involves soft skills such as decorating things, then certain groups of students (academically/scientifically inclined) are less likely to consider design as important for their career. It would also be difficult for them to engage in dialogue on design or evaluate everyday products on design grounds. Secondly, in my study, most of the design activities were related to handling or manipulating of physical artefacts. Even the design problem given to the students entailed designing and making a tangible model/product. However, it would be good to know how students' ideas of design change if there is an increase in the range of different problem-solving contexts from the tangibles to intangibles.

How Teachers Might Contribute to These Investigations

Firstly, teachers who desire to know what ideas and images their students associate with design and designers may use the survey questionnaire. Even a simple draw-a-designer-at-work task would provide insights into students' understanding of the nature of design and design professions. Secondly, besides engaging students in the manipulation and designing of physical products, a teacher can choose a range of other problem-solving contexts which involve the designing of intangibles such as services/experiences, say, traffic signals, experiences of people in the hospital waiting rooms, at the railway stations, or designing a drama including creating stories, characterisation and staging a story. These problems would not only enrich students' understanding of design but also provide them opportunities to understand complexity in design problems, human behaviour and their emotional and psychological needs. These problems may not rely heavily on the skills in drawing, manipulation of materials or actual making, but involve skills of observation, storytelling and acquaintance with new technology and media.

Acknowledgement I am grateful to Prof. John Williams and Dr. David Barlex and other reviewers for going through all my drafts and providing suggestions and feedback while I was writing this chapter. I would also like to thank Prof. Sugra Chunawala, the supervisor of my thesis for her feedback and constant support during my PhD years.

References

Anning, A., Jenkins, E., & Whitelaw, S. (1996). Bodies of knowledge and design-based activities. In J. S. Smith (Ed.), *IDATER 1996*. Loughborough: Loughborough University.

Ara, F., Chunawala, S., & Natarajan, C. (2009). From analysing to designing artefacts: Studying middle school students' ideas about design and designers. In K. Subramaniam, & A. Mazumdar (Eds.), *Proceedings of epiSTEME-3: International conference to review research in science, technology and mathematics education*. Mumbai: Macmillan

Ara, F., Chunawala, S., & Natarajan, C. (2011). A study investigating Indian middle school students' ideas of design and designers. *Design and Technology Education: An International Journal, 16*(3), 62–73. http://ojs.lboro.ac.uk/ojs/index.php/DATE/article/view/1676/1571

Ara, F., Chunawala, S., & Natarajan, C. (2013). Investigating Indian elementary and middle school students' images of designers. *Design and Technology Education: An International Journal, 18*(2), 50–65. https://ojs.lboro.ac.uk/ojs/index.php/DATE/article/view/1839/1756

Atkinson, S. (2009). Are design and technology teachers able to meet the challenges inherent in the theme for this conference 'D&T—A platform for success'? In E. Norman (Ed.), *Design and Technology Education: International Journal, 14*(3), 8–20.

Barlex, D., & Rutland, M. (2004). Design decisions in Nuffield design & technology. In I. Mottier, & M. de Vries (Eds.), *Proceedings of 14th PATT conference*. New Mexico, USA.

Bronowski, J. (1973). *The ascent of man*. Boston: Little, Brown and Company.

Butler, C. (2012). *Your ego is a bad designer*. Retrieved August 29, 2012, from Newfangled Web Developers: http://www.newfangled.com/your_ego_is_a_bad_designer

Chambers, D. W. (1983). Stereotypic images of the scientist: The draw-a-scientist test. *Science Education, 67*(2), 255–265.

Choksi, B., Chunawala, S., & Natarajan, C. (2006). Technology education as a school subject in the Indian context. In K. Volk (Ed.), *Articulating technology education in a global community: International conference on technology education in the Asia Pacific region*. Hong Kong: Hong Kong Technology Education Association and The Hong Kong Polytechnic University.

Fralick, B., Kearn, J., Thompson, S., & Lyons, S. (2009). How middle schoolers draw engineers and scientists. *Journal of Science Education and Technology, 18*, 60–73.

Glick, P., Wilk, K., & Perreault, M. (1995). Images of occupations: Components of gender and status in occupational stereotypes. *Sex Roles, 32*, 564–582.

Hennessy, S., & McCormick, R. (2002). The general problem-solving process in technology education: Myth or reality. In G. Owen-Jackson (Ed.), *Aspects of teaching secondary design and technology: Perspectives on practice*. London: RoutledgeFalmer.

Jarvis, T., & Rennie, L. (1998). Factors that influence children's developing perceptions of technology. *International Journal of Technology and Design Education, 8*, 261–279.
Jones, A., & Carr, M. (1994). Student technological capability: Where do we start? *SAMEpapers, 1994*, 165–186.
Khunyakari, R., Mehrotra, S., Chunawala, S., & Natarajan, C. (2009). Studying Indian middle school students' attitudes towards technology. In K. Subramaniam, & A. Mazumdar (Eds.), *Proceedings epiSTEME-3: An international conference to review research in science, technology and mathematics education* (pp. 81–87). Mumbai: Macmillan Publishers India Ltd.
Kimbell, R., & Perry, D. (2001). *Design and technology in a knowledge economy*. London: Engineering Council.
Knight, M., & Cunningham, C. (2004). Draw an Engineer Test (DAET): Development of a tool to investigate students' ideas about engineers and engineering. *Proceedings of the 2004 American Society for Engineering Education annual conference & exposition.* American Society for Engineering Education.
Layton, D. (1994). A school subject in the making? The search for fundamentals. In D. Layton (Ed.), *Innovations in science and technology education* (pp. 11–28). France: UNESCO Publishing.
Lewis, T., Petrina, S., & Hill, A. (1998). Problem posing-adding a creative increment to technological problem solving. *Journal of Industrial Teacher Education, 36*(1), 5–35.
Mawson, B. (2007). Factors affecting learning in technology in the early years at school. *International Journal of Technology and Design Education, 17*(3), 253–269.
Mead, M., & Metraux, R. (1957). Image of the scientist among high school students: A pilot study. *Science, 126*, 386–390.
Mehrotra, S., Khunyakari, R., Chunawala, S., & Natarajan, C. (2007). Using pictures and interviews to elicit Indian students' understanding of technology. In J. D. Dakers, & M. de Vries (Eds.), *Proceedings of PATT 18 – International conference on design and technology education research: teaching and learning technological literacy in the classroom* (pp. 152–161).
Newstetter, W., & McCracken, M. (2001). Novice conceptions of design: Implications for the design of learning environments. In C. M. Eastman, W. M. McCracken, & W. C. Newstetter (Eds.), *Design knowing and learning: Cognition in design education* (pp. 63–77). Oxford: Elsevier Science Ltd.
Owen, C. L. (2005). *Design thinking. What it is. Why it is different. Where it has new value.* Keynote speech at the international conference on design research and education for the future. Institute of Design, Illinois Institute of Technology.
Owen-Jackson, G. (2013). *Debates in design and technology education.* Abingdon: Routledge.
Picker, S., & Berry, J. (2000). Investigating pupils' images of mathematicians. *Educational Studies in Mathematics, 43*(1), 65–94.
Roberts, P. (2005). Of models, modelling and design: An applied philosophical enquiry. In L. Archer, K. Baynes, & P. Roberts (Eds.), *A framework for design and design education: A reader containing papers from the 1970s and 80s*. Wellesbourne: D&T Association.
Rogers, M., & Clare, D. (1994). The process diary: Developing capability within national curriculum design and technology- Some initial findings. In J. S. Smith (Ed.), *IDATER 94: International conference on design and technology educational research and curriculum development,* (pp. 22–28). Loughborough University of Technology.
Welch, M., Barlex, D., & O'Donnell, E. (2006). Elementary students' beliefs about designers and designing. In E. W. Norman, D. Spendlove, & G. O. Jackson (Eds.), *Designing the future: The D&T Association international research conference*, (pp. 165–176).

For Further Reading

An electronic copy of the PhD thesis can be found at this url:http://www.hbcse.tifr.res.in/research-development/ph.d.-theses/farhat-ara.pdf

Banks, F. (1994). *Teaching technology*. London: Routledge.

Cross, N. (2006). *Designerly ways of knowing*. London: Springer.

Dakers, J. R. (2006). *Defining technological literacy: Towards an epistemological framework*. New York: Palgrave Macmillan.

Johnsey, R. (1995). The design process – Does it exist? *International Journal of Design and Technology Education, 5*(3), 199–217.

Lawson, B. (2005). *How designers think: The design process demystified*. Oxford: Architectural Press.

Stables, K. (2008). Designing matters; Designing minds: The importance of nurturing the designerly in young people. *Design and Technology Education: An International Journal, 13*(3), 8–18.

Trebell, D. (2009). Exploring pupils' beliefs about designers and designing. *International Journal of Technology and Design Education, 21*(1), 19–54.

Chapter 3
Fostering Deep Learning and Critical Thinking Amongst Net Generation Learners

Matthew Watkins

Abstract This chapter explores how to encourage Net Generation learners to develop deep learning through the use of carefully designed interactive audio-visual workshops. Careful design of teaching resources can increase relevance, encourage discussion and debate and provide opportunities for critical reflection amongst students, a noted weakness of Net Generation learners.

The Questions I Asked and Why I Think They Are Important

How to encourage students to foster deep learning? This was considered important as the literature suggested that deep learning is crucial in helping students to grasp the holistic nature of the social aspects of sustainability (Warburton 2003).

> Deep learning involves paying attention to underlying meaning. It is associated with the use of analytical skills, cross-referencing, imaginative reconstruction and independent thinking (Warburton 2003)

The literature states that there need to be the correct conditions in place to foster deep learning. These conditions include ensuring relevance (Ramsden 1997; McMahon 2006) and motivation (Fransson 1977; Hounsell 1997; Marton and Säljö 1997) and reducing anxiety, often born out of the threat of assessment (Fransson 1977; McMahon 2006); therefore the research considered the learning preferences of the students to address these factors. The current generation of students can be known as 'Net Generation' learners (Oblinger and Oblinger 2005b), Millennials (Howe and Strauss 2003; Holliday and Li 2004) or 'digital natives' (Tapscott 2009; Palfrey and Gasser 2008) and are typically designated as those born after 1982. However Oblinger and Oblinger (2005a) note that the differentiating factor for the 'Net Generation' may be their technological experience rather than just their age,

M. Watkins (✉)
School of Architecture, Design and the Built Environment, Nottingham Trent University, Nottingham, UK
e-mail: matthew.watkins@ntu.ac.uk

PJ. Williams, D. Barlex (eds.), *Contemporary Research in Technology Education*, Contemporary Issues in Technology Education, DOI 10.1007/978-981-10-2819-9_3

with such students having grown up with the use of the Internet and a rich digital media environment. Due to this, the Net Generation have differing learning styles and preferences to previous generations of learners (Barnes et al. 2007), which commonly includes many of their educators and instructors.

Literature on the Net Generation learning preferences was considered and found that this generation of students are autonomous learners who place a greater emphasis on exploratory learning through discovery (Oblinger and Oblinger 2005a; Tapscott 2009), which aids their ability to retain knowledge (Tapscott 2009). Net Generation learners also have a preference for group-based inquiry (Howe and Strauss 2003; Tapscott 2009; Barnes et al. 2007; Oblinger and Oblinger 2005a) as they are very social learners who prefer a peer-to-peer (Oblinger and Oblinger 2005a) and interactive approach (Tapscott 2009; Barnes et al. 2007). They are also described as visual learners (Holliday and Li 2004) with enhanced visuospatial skills (Tapscott 2009; Oblinger and Oblinger 2005a), with the literature suggesting that Net Generation learners are more comfortable with image-rich environments than with text. Typically they retain 30 % of what they see but only 10 % of what they read (Oblinger and Oblinger 2005a; Tapscott 2009; Windham 2005) and prefer to have graphics before text rather than graphics following text (Oblinger and Oblinger 2005a). Oblinger and Oblinger (2005a) note that Net Generation learners have a highly developed visual literacy, with the ability to read images and instinctively communicate through visual methods. They are also capable of combining images, text and sound seamlessly (Oblinger and Oblinger 2005a), which is demonstrated by the prevalence of amateur You Tube content.

However, the Net Generation lack the ability to reflect and adopt critical thinking skills (Holliday and Li 2004; Oblinger and Oblinger 2005b) due to their propensity to multitask. Therefore in order to foster deep learning, the weaknesses of the Net Generation need to be overcome and the strengths and preferences of Net Generation learners exploited.

Therefore the research considered how to encourage critical reflection, discussion and debate amongst learners in a group context, by addressing the weakness of such students and encouraging them to go beyond a surface level of learning, which is consistent with repeating what they had seen without evaluation.

How I Tried to Answer the Questions

A multidisciplinary literature review was conducted which considered the learning preferences of the Net Generation, how to foster deep learning and the use of audio-visual resources, encompassing fields as diverse as social sciences, zoology, education and design.

Due to a visual preference and the growing prevalence of an audio-visual culture through formats such as YouTube amongst the student generation, audio-visual (A/V) learning approaches were explored as a medium. However, a lack of literature existed within design education on the benefits of such an approach, so literature was considered from the field of sociology education, where the practice of using both photographs and music is well established, particularly in relation to teaching introductory sociology. This literature was closely aligned with the research, which sought to develop an introductory approach for teaching the social aspects of sustainability within product design.

Literature suggests that an A/V approach can have reflective benefits (Albers and Bach 2003; Tan and Pearce 2011) and foster critical thinking and reflection, by encouraging discussion in a group context through the creation of a shared experience (Albers and Bach 2003; Hraba et al. 1980). Such benefits are crucial as they address the key perceived weakness of Net Generation learners (Holliday and Li 2004; Oblinger and Oblinger 2005a). Additionally a visual environment suits the preferences of the Net Generation, whilst the additional use of audio in the form of music is noted in the literature as beneficial to learning (Ahlkvist 2001; Albers and Bach 2003; Brkich 2012).

Three A/V workshops were developed, each consisting of two elements, a 3–5 min A/V introduction featuring a variety of photographs, which were overlaid with popular music related to each topic and a 45 min group-based workshop session to respond to the A/V introduction. The workshops were designed to address three overarching aspects of social sustainability:

- Design for the other 90 % (Watkins 2014c)
- Localisation, emotionally durable design and reuse (Watkins 2014b)
- Exploitation (Watkins 2014a)

However, the approach used is universal and could be applied to a variety of topics. To complement the A/V introductions, four prompting questions were devised to encourage the students to reflect on the content and respond. The use of questioning techniques is recognised in both the literature (Griffith and Bamford 2007) and earlier academic interviews in the PhD study as beneficial to critical thinking and group discussion.

The workshops were each self-contained units and therefore can be run in a flexible manner and in different combinations depending on the need and availability. Within the research, all three were conducted back to back in 1 day at one institution, whilst at another institution, two workshops were spread across 2 consecutive days, and two were run across a number of weeks at several other institutions, whilst at one university, a single workshop was run as a one-off.

The A/V introductions were shown at the start of each workshop session, and students were asked to respond to each individually before being placed in small groups of 4–5 students. The groups were asked to discuss each member's responses before the prompt questions were given. Typically, students were given 10 min to discuss each question and document their responses on A1 flipchart paper in the form of a 'thought shower', which is a group-based spontaneous problem-solving exercise often in the form of a spider diagram. Brainstorming and reflection are cited as important to the design thinking process (Seidel and Fixson 2013).

The photographs used in the introductions were selected intentionally to portray a number of different aspects, echoing the well-known Chinese proverb, 'one picture is worth ten thousand words', so that each A/V introduction could introduce a much larger range of social issues than a traditional lecture format could accommodate, if only at a superficial level. Photographs were also chosen as the literature findings suggest that the use of images can elicit an indirect experience which can foster personalisation of sustainability (Murray 2011). Small groups were designated to enable students to engage with discussion, debate and critical reflection as well as encouraging engagement (McNerney and Davis 1996; Huckle and Sterling 1997).

What I Found Out

Data from the workshops was collected using a number of methods; audio recording and still photography were used to document the workshop, whilst identical questionnaires were conducted prior to the start of the workshops and again upon completion to measure changes in knowledge and attitude. In addition to this, reflective diaries that the students kept were examined for references to the workshops. The data was subjected to coding and clustering analysis (Miles and Huberman 1994) in order to maintain the integrity of the data and clearly identify themes.

Relevance of the A/V Introductions

The students responded very positively to the design of the workshops and particularly engaged with the audio-visual introductions, describing the audio-visual introductions as effective and informative, noting the short length, simplicity and thought-provoking nature as well as the ability to evoke discussion. Students described how the audio-visual format was memorable and helped them to think, citing the emotive nature of the music. Descriptions of the memorable and provocative nature of the A/V introductions included terminology such as 'powerful' 2019, 'striking' and even 'upsetting' or 'shocking' in response to the exploitation of workers.

The inclusion of the music was found to be effective. This was evidenced by students discussing the nature of the lyrics of a song used in one of the introductions, which suggest stepping into someone else's world. The students effectively applied this to the diversity of the individuals portrayed in the A/V introduction, leading to a variety of empathic user-centred design research approaches being suggested. 'Lyrics of the song – take a step into my world – suggest living in one of these peoples shoes for a day'.

The use of images and the visual nature of the A/V introductions were found to enhance the memorable nature of the introductions with students later identifying individual images, objects or individuals portrayed and describing the 'striking' nature of the photographs and composition: 'It was the wonderful selection of photos that made me stop and think'.

Students also began to relate to the examples given in the A/V introduction personally, contextualising aspects with their own experience. Examples included students describing objects that they have an affinity with, recognising aspects of socially orientated sustainable design such as emotionally durable design, personal meaning and reuse.

In addition to citing the images, students also recalled text-based quotes used in the A/V introductions, such as 'One man's waste is another man's treasure', which was interesting and relates to how the use of text following graphics is more effective (Oblinger and Oblinger 2005a).

Effectiveness of the Workshop Exercises in Fostering Deep Learning

The literature review demonstrated a link between deep learning, motivation, collaboration and critical reflection; therefore, the following section will also consider how these attitudes featured in the analysis of the student workshop experience.

Students described the workshop exercises as enjoyable and valuable, citing how the workshops had helped them broaden their thinking and outlook in respect to their design solutions.

The students described how working in groups was beneficial to their learning and helped them to recognise different perspectives, through the wider views of others. By reflecting on the conflicting views that arose within the group discussion, students recognised that there were often two equally valid viewpoints or arguments to a particular issue: grasping an important characteristic of the complexity of sustainability and demonstrating critical reflection. One student in particular described his preference for group work, echoing the Net Generation literature which suggests that the current generation of students find a peer-to-peer learning approach more credible than a teacher learning experience.

The workshops appeared to foster changes in the students' attitudes towards ethics in design. Following the workshops, students prioritised the values and ethical considerations in design more highly in relation to more practical and tangible design considerations, which were measured by identical questionnaires conducted prior to and following the workshops and from the students' own reflections through their design journals.

These reflections extracted from their reflective diaries demonstrated how they had personalised the impact of the workshops, questioning both how it affected their practice and responsibilities as designers as well as their responsibilities as consumers and citizens.

> It showed me that as a designer I have an obligation to use the talent and career that I have been given to help others.

However, such reflection didn't occur automatically; rather, it was progressive and developed from the students' interaction both personally and communally throughout the workshops. Initial individual responses to the A/V introductions taken after the first 5 min were found to be predominately surface level learning and merely observations and recall of what they had seen. However, the students' ability to reflect critically upon what they had seen began to develop through the group phase of the workshop alongside the discussion and debate that followed with their peers.

This differentiation between the individual responses and group-based sessions would suggest that group work, group discussion and mind mapping were the key aspects in encouraging students to engage with the material at a deeper level and begin the transition from surface learning to deep learning, supporting the design of the workshops and literature findings by encouraging critical reflection (Vaughan 2006; McMahon 2006; Entwistle 2000; Gokhale 1995), collaborative learning (Entwistle 2000; McMahon 2006; Vaughan 2006), learning by discovery (Warburton 2003) and enabling students to internalise their learning through sharing with others (Tapscott 2009).

This led students to engage with aspects at a personal level as well as engaging in critical reflection with each other, discussing the designer's responsibilities and, for example, the complexity of child labour in developing nations and suggesting approaches which fostered design thinking principles. Evidencing analytical and independent thinking skills concerned with the underlying issues of what was presented echoed attributes that Warburton (2003) describes as deep learning.

Throughout all three workshops, students endeavoured to relate the workshop content to design, considering how design approaches could address the issues raised. In particular, they discussed design for need and how design for all isn't always possible, discussing instead designing for groups of people and identifying cultural differences relating to age, ability or ethnic origin. Students also differentiated between the need to design for 'needs as opposed to merely wants'.

Seeking to address some of the more complex global issues that had arisen in the workshops through design was a challenging process for the students that lead them to suggest innovative design thinking approaches, to solve issues that typically exist beyond the traditional remit of the designer. For example, from one student:

> We first looked at the Irish beef market where you can trace meat from farm to fork and thought the same process could be employed in the manufacturing industry each material and component could be traced back to a country and company of origin making every link visible in the chain.

This evidenced the ability to cross-reference and imaginatively think out of the box applying existing knowledge to new contexts: attributes of both deep learning and the emerging field of design thinking. Students continued to reflect upon what they had learnt following the workshops in their journals, unravelling issues that underlined some of the issues explored within the workshops, beginning to question macro themes such as globalisation, mass production and the rapid advancement of technology.

It was found that even the dual questionnaire method of data collection encouraged reflection amongst the students who later discussed in their journals how they had reflected upon what they had learnt whilst completing the second questionnaire and considering how their opinions had changed since they conducted the first questionnaire prior to the workshops. For example:

> After the workshop we were then asked to fill in the same survey as we did before the workshop. This was interesting because my thoughts and knowledge of the aspects had changed or broadened.

Students also began to carry across learning or content from one workshop into another building upon prior learning and demonstrating the ability to cross-reference an important characteristic of deep learning.

Application of Learning from Workshops

There were examples of students later applying learning from the workshops through their practical design activities. One student group explored the theme of intergenerational design that they arrived at during the 'design for the other 90 %' workshop, producing a concept for multifunctional public seating that encouraged different generations to interact with each other, affirming the importance of family. This built upon the discussions and conclusions that the group had come too jointly whilst reflecting on the needs of different generations and the difficulties of broken families, disaffected youth and loneliness amongst the elderly.

How this Might Be Used to Improve Teaching and Learning

The immediate benefits of such A/V resources and the styles of teaching discussed within this chapter are transferable from higher education to a secondary level. The workshops were designed in such a way as to enable them to be replicated by others and even be taught by non-specialists in sustainability, due to their introductory nature and ability to enable students to question the themes themselves and engage in autonomous learning. Questions concerning the potential for delivery by non-specialist staff in sustainability were addressed in the original research by interviewing academics involved within the institutions trialled.

Academics agreed that the materials could be used by non-specialists but that difficulties may be encountered should the students ask direct questions of staff during or after the exercise and suggested that introductory academic guidance is also given with additional information on specific images to help academics deliver the material autonomously in their own institutions. As one student stated:

> Yes but I think the lecturer will need to have some understanding or maybe some notes or a recommended reading list to support the presentations with additional examples so that they can use them to facilitate discussion with the students.

It was also hoped that such workshops would be able to be reproduced by others following the principles outlined in the research to create introductory A/V introductions and questions on a number of different themes to suit their individual teaching. Whilst this is possible, the researcher recognises that producing the A/V introductions requires skill sets that may not be common to all teachers such as an in-depth knowledge of enough contemporary music to be able to identify suitable audio, or a visual 'eye' to select photographs and images, which convey additional meaning. However there are a number of Internet-based tools available to assist educators, such as Panos (Panos Pictures 2012), a network of documentary photographers that specialise in global social issues photography. Furthermore, there are numerous sites which document the lyrics to contemporary music as well as sites that contain lists of songs that deal with global issues and social justice with songs for use in education, such as Jam for Justice (Cohen 2014), rethinking schools (Peterson 2002), EdChange (Gorski 2014) and SoJust (2012).

Beyond the design and development of the A/V introductions, the workshops' format also requires careful planning and objectivity to avoid bias and ensure that students are able to develop their own critical understanding of the issues involved. This was ensured within the research by identifying stereotypes and then carefully selecting images that don't conform to stereotypical depictions. An example of this was the intentional use of an image of an outgoing young man in a wheelchair shown in the 'Step into my World' A/V introduction. The group work element of the workshops is also important, enabling students to consider multiple perspectives and critically reflect on what they have seen.

Additional care needs to be taken with the prompt questions used within the workshops to ensure that questions are not leading but instead encourage the students to think critically and form their own perspectives on the materials in order to foster deep learning. With younger groups of students, it may be necessary for the teacher to give further prompts by asking more open questions relating to the content in their thought showers in order to enhance the engagement of all learners.

Addressing Key Content

One of the most relevant transferable elements within these workshops is the content itself, with sustainability considerations now commonplace in the design and technology curricula in the United Kingdom (DATA 2014) and New Zealand (Ministry of Education 2007) and technology standards in the United States (ITEA 2007). All of these contain statements which relate the students' ability to consider and responsibly use technology in relation to society and the environment. The workshops already developed address a wide range of areas including the following:

- Range of domestic, local and industrial contexts – through a consideration of localisation and wider issues arising from globalised production

- User needs – considering the needs of users and diversity of user needs
- Different cultures – considering user needs and cultures that are unfamiliar
- View of intended users – considering the intended end users of their products
- Impact – both in relation to the environmental and social impacts of sustainability including globalisation

The following are extracts taken from either the UK National Curriculum (DATA 2014), the New Zealand Curriculum (Ministry of Education 2007) or the US Standards for Technological Literacy (ITEA 2007), to highlight the considerations of sustainability in each:

- Understand the role that designers and product developers have and the impact and responsibility they have on and to society (DATA 2014).
- Learn to consider ethics, legal requirements, protocols, codes of practice and the needs of and potential impacts on stakeholders and the environment (Ministry of Education 2007).
- Understand that designers and manufacturers have both a moral and legal responsibility for the products that they create (DATA 2014).
- Consider environmental and sustainability issues in designing products (DATA 2014).
- Understanding the lifecycle of a material or product and be able to trace the lifecycle of a product from its inception to its disposal (ITEA 2007).
- Take into consideration the ethical, environmental and sustainability issues relating to the design and manufacture of products, i.e. fair trade, product miles, carbon footprint, product disposal and the following related principles: reuse, recycle, repair, reduce, rethink, refuse, etc. (DATA 2014).
- Students should explore how a waste product is recycled, reused or remanufactured into a new product (ITEA 2007).
- Have a knowledge and understanding of the main factors relating to recycling and/or reusing materials or products, i.e. material identification, material separation, collection, processing, energy costs, subsequent usage and wastage (DATA 2014).
- Students should research the potential clash between environmental and economic concerns created by technological products and systems (ITEA 2007).
- Exploring the long-term impact of social, cultural, scientific, technological, economic or political practices on society and the environment (Ministry of Education 2007).
- Learn to critique the impact of technology on societies and the environment (Ministry of Education 2007).
- Students need to understand the delicate balance amongst people, technology and the environment (ITEA 2007).
- Consider the conflicting demands that moral, cultural, economic and social values and needs can make in the planning and in the designing of products (DATA 2014).
- Exploring what it means to be a citizen and to contribute to the development and well-being of society (Ministry of Education 2007).

- Understand social, economic and ethnic groups of people often have specific values and needs which can be an aid to focused designing, i.e. disabled, elderly and religious groups (DATA 2014).

The nature of the content of the workshops also enables teachers to draw parallels with aspects of citizenship curricula within design and technology, in relation to the wider social and ethical considerations of design and manufacture, enabling students to recognise transferable content and ensure a real-world focus that is consistent with enhancing the relevance of taught content. This was supported by students' reflections to the workshops, which noted that it changed their attitudes not only as designers but also as consumers and through the impact that they can have on others.

> It showed me that as a designer I have an obligation to use the talent and career that I have been given to help others.
>
> I thought about how a simple design or idea can have a big impact. Not on a consumer of an item, but on a community or a way of life. It made me really think about the power of design, as well as the responsibilities of designers to stay relevant and that as well as being fashionable or popular, that design has an important message, or in this case an important cause.

Delivery Methods

In addition to the content of the workshops, the teaching and learning principles underlying them to foster deep learning and critical thinking are applicable to students at different levels.

- *Audio-visual introductions*

The thought-provoking nature of the A/V introductions and associated workshop questions led the students to consider a range of criteria and such resources can be used as a catalyst to encourage students to begin researching complex areas such as sustainability and explore key areas within the curriculum in a manner that encourages independent learning. The benefit of such resources within the research was that such focussed workshops can be used by students as a starting point for developing and writing their own design briefs.

- *Students developing their own briefs*

Following the completion of the doctoral research, the researcher has continued to use the Rethinking Design workshops in his teaching, encouraging second-year undergraduate students to derive their own briefs and intended outcomes from the experience. Whilst this has been applicable to undergraduate students, younger students may require further guidance in order to fulfil the task satisfactorily perhaps with a guidance sheet and directed questions. This principle of encouraging students to set their own assessment criteria is consistent with assessment methods that are more likely to foster deep learning, through reduced anxiety and greater motivation and exploratory learning.

Development of such design briefs may help to develop more worthy real-world projects with detailed research being considered by 14–18-year-old students in relation to coursework projects. For example, the 'Rethinking Design' workshops could encourage students to consider wider sustainability concerns such as the environment and society, considering the impact upon the end users and stakeholders affected throughout the manufacture and processing of such products.

Outputs from such open briefs derived from the 'Step into my World' workshop have resulted in second year undergraduate students coming up with a variety of proposals including a social network community-based solution to address loneliness amongst the elderly, a product that ex-military personnel are given at their end of service to help them keep in touch with their peers and address issues such as post-traumatic stress disorder and an item of intergenerational public seating that encourages different generations to communicate with each other.

- *Reflective journals*

Another applicable method of enhancing students' learning arising from the 'Rethinking Design' workshops is encouraging the use of reflective journals amongst students, which is recognised also within education literature (McMahon 2006). Students can be asked to use reflective journals to reflect upon their learning; particularly aspects that were memorable, inspiring or challenging. Such an approach fosters deep learning through reflection whilst also enabling the educators to recognise which aspects of teaching were most successful, further assisting the development of effective teaching strategies.

Furthermore, encouraging students to reflect upon their learning through writing will help them to digest their learning and foster the designerly thinking patterns that lead to the 'incubation of ideas' through subconscious reflection as noted in design education literature (Glegg 1969; Whitfield 1975; Lawson 2006).

What Might Be Investigated Further

In addition to the research already conducted, the following considerations for further research could help to support the fostering of deep learning and critical thinking amongst Net Generation learners in design and technology subject areas.

Effectiveness of the Approach Beyond the Net Generation

Future learners are likely to become even more familiar with technology and the Internet from an even earlier age with the associated benefits and limitations that this will bring in terms of their learning and ability to engage in deep learning and critical thinking. Therefore, the beneficial impact of the A/V interactive approaches documented within this chapter may change as younger learners move through the education system, and therefore the continued effectiveness of the teaching approach should be monitored to ensure continued relevance.

Upstream Effects of Enhancing Learning

How will fostering independent learning, deep learning and critical reflection amongst younger learners affect their future capacity for learning prior to higher/tertiary education? Could such methods potentially decrease the learning curve experienced by students moving between educational levels?

Will future students entering higher education be more capable of independent learning and more familiar with the principles of sustainability as a result? And if so, how will this affect the curricula and capacity within higher education? It may be beneficial for educators to undertake audits of a student's capacity to engage in deep, critical and independent in order to inform future levels of study.

How Teachers Might Contribute to These Investigations

How could the resources described within this chapter be adapted to accommodate younger learners, specifically 11–14-year-old students that are more likely to be challenged by the open-ended nature of the resources and requirement for independent learning? Further studies by educators are required with younger age learners using the approaches explored within this chapter, with particular attention paid to the student's experience, through student feedback and review of the outputs. Additionally, it would be very interesting to see how the resources affect students' future learning and understanding through a longitudinal study across the 11–18 age range, which is more feasible over a longer time frame within secondary level education than can be undertaken in higher/tertiary education, which is limited to the length of study and modular content.

Can opportunities for students to create their own design briefs be managed in such a way that the depth, complexity and holistic nature of sustainability aren't diminished? Perhaps educators could provide or be provided with sustainability guidance to help students recognise the complexities and inherent conflicts when addressing the three pillars of sustainability ensuring that solutions are equally environmentally benign, economically viable and socially equitable (Elkington 1998). An example of this was explored in one of the three workshops noted earlier, which considered the disadvantages of globalisation in terms of the environment impact and social impact through exploitation of workers, but was contrasted with the need for employment within the developing economies.

Dissemination from educators reporting on trails amongst 11–14- and 14–18-year-old students would be very beneficial to the furthering of this research, particularly if they documented what further guidance was required and implemented.

An electronic copy of the PhD Thesis can be found in the British Library; http://ethos.bl.uk/OrderDetails.do?uin=uk.bl.ethos.594455

References

Ahlkvist, J. A. (2001). Sound and vision: Using progressive rock to teach social theory. *Teaching Sociology, 29*(4), 471–482.

Albers, B. D., & Bach, R. (2003). Rockin' soc: Using popular music to introduce sociological concepts. *Teaching Sociology, 31*(2), 237–245.

Barnes, K., Marateo, R., & Ferris, S. (2007). Teaching and learning with the net generation. *Innovate, 3*(4).

Cohen, S. B. (2014). *Songs for justice*. Retrieved August 18, 2011, from http://www.jamforjustice.org/the-experience/songs-for-justice/

DATA. (2014). *Design and technology programmes of study: Key stages 1 to 3 national curriculum in England key*. Wellesbourne, Warwickshire. Retrieved May 30, 2015, from https://www.data.org.uk/for-education/curriculum/dt-national-curriculum-for-england-2014/

Elkington, J. (1998). *Cannibals with forks: the triple bottom line of 21st century business*. Gabriola Island: New Society Publishers.

Entwistle, N. J. (2000). Promoting deep learning through teaching and assessment: Conceptual frameworks and educational contexts. In *TLRP conference*. Leicester. Retrieved from http://www.etl.tla.ed.ac.uk//docs/entwistle2000.pdf

Fransson, A. (1977). On qualitative differences in learning: IV – Effects of intrinsic motivation and extrinsic test anxiety on process and outcome. *British Journal of Educational Psychology, 47*, 244–257.

Glegg, G. L. (1969). *The design of design*. London: Cambridge University Press.

Gokhale, A. A. (1995). Collaborative learning enhances critical thinking. *Journal of Technology Education, 7*(1), 22–30.

Gorski, P. C. (2014). *Social justice song index*. Retrieved May 30, 2015, from http://www.edchange.org/multicultural/arts/songs.html

Griffith J. S., & Bamford, R. (2007). *Embedding the principles of responsible design into student practice from 1st year*. In: Proceedings of connected 2007 international conference on design education. University of New South Wales, Sydney, Australia.

Holliday, W., & Li, Q. (2004). Understanding the millennials: Updating our knowledge about students. *Reference Services Review, 32*(4), 356–366.

Hounsell, D. (1997). Understanding and teaching for understanding. In N. Marton, F. Hounsell, & D. Entwistle (Eds.), *The experience of learning: Implications for teaching and studying in higher education* (pp. 238–257). Edinburgh: Scottish Academic Press.

Howe, N., & Strauss, W. (2003). *Millenials go to college: Strategies for a new generation on campus: Recruiting and admissions, campus life, and the classroom* (2nd ed.). Washington, DC: American Association of Collegiate Registrars and Admissions Officers.

Hraba, J., Powers, E., Woodman, W., & Miller, M. (1980). Social change through photographs and music: A qualitative method for teaching. *Qualitative Sociology, 3*(2), 123–135.

Huckle, J., & Sterling, S. (1997). *Education for sustainability*. London: Earthscan.

ITEA. (2007). *Standards for technological literacy: Content for the study of technology*. Reston: International Technology Education Association. Retrieved May 30, 2015, from http://www.iteea.org/TAA/Publications/TAA_Publications.html

Lawson, B. (2006). *How designers think: The design process demystified* (4th ed.). Oxford: Architectural Press.

Marton, F., & Säljö, R. (1997). Approaches to learning. In F. Marton, D. Hounsell, & N. J. Entwistle (Eds.), *The experience of learning: Implications for teaching and studying in higher education* (pp. 39–58). Edinburgh: Scottish Academic Press.

McMahon, T. (2006). Teaching for more effective learning: Seven maxims for practice. *Radiography, 12*(1), 34–44.

McNerney, C., & Davis, N. D. (1996). *Education for sustainability: An agenda for action*. Pennsylvania: Diane Publishing Company.

Miles, M. B., & Huberman, A. M. (1994). *Qualitative data analysis: An expanded sourcebook* (2nd ed.). Thousand Oaks: Sage.

Ministry of Education. (2007). *The New Zealand curriculum*. Wellington: Ministry of Education.

Murray, P. (2011). *The sustainable self: A personal approach to sustainability education*. London: Earthscan.

Oblinger, D., & Oblinger, J. (2005a). *Educating the net generation*. Retrieved August 18, 2011, from http://www.educause.edu/educatingthenetgen

Oblinger, D., & Oblinger, J. (2005b). *Educating the net generation – Is it age or IT: First steps toward understanding the net generation*. Retrieved August 18, 2011, from http://www.educause.edu/educatingthenetgen

Palfrey, J., & Gasser, U. (2008). *Born digital: Understanding the first generation of digital natives*. New York: Basic Books.

Panos Pictures. (2012). No Title. Retrieved September 20, 2012, from http://www.panos.co.uk/

Peterson, B. (2002). *Songs with a global conscience*. Retrieved May 30, 2015, from http://www.rethinkingschools.org/publication/rg/RGResource01.shtml

Ramsden, P. (1997). The context of learning in academic departments. In N. Marton, F. Hounsell, & D. Entwistle (Eds.), *The experience of learning: Implications for teaching and studying in higher education* (pp. 198–216). Edinburgh: Scottish Academic Press.

Seidel, V. P., & Fixson, S. K. (2013). Adopting "design thinking" in novice multidisciplinary teams: The application and linits of design methods and reflexive practices. *Journal of Product Innovation Management, 30*(S1), 19–33.

So Just. (2012). *Social justice songs*. Retrieved May 30, 2015, from http://www.sojust.net/songs.html

Tan, E., & Pearce, N. (2011). Open education videos in the classroom:exploring the opportunities and barriers to the use of youTube in teaching introductory sociology. *Research in Learning Technology, 19*(1), 125–133.

Tapscott, D. (2009). *Grown up digital: How the net generation is changing your world* (1st ed.). New York: McGraw-Hill.

Vaughan, N. (2006). The use of wikis and weblogs to support deep Approaches to learning (Vol. 1). Calgary: Teaching and Learning Centre, University of Calgary. Retrieved from http://journals.ufv.ca/rr/RR13/article-PDFs/6-vaughan.pdf

Warburton, K. (2003). Deep learning and education for sustainability. *International Journal of Sustainability in Higher Education, 4*(1), 44–56.

Watkins, M. A. (2014a). *Exploitation – Rethinking design series*. Retrieved May 30, 2015, from http://youtu.be/D7LRY7bEbyw

Watkins, M. A. (2014b). *Localisation, emotionally durable design and reuse – Rethinking design series*. Retrieved May 30, 2015, from http://youtu.be/9RKSl6-Ewcc

Watkins, M. A. (2014c). *Step into my world – Rethinking design series*. Retrieved May 30, 2015, from http://youtu.be/FtAZv1l7_80

Whitfield, P. R. (1975). *Creativity in industry*. Harmondsworth: Penguin.

Windham, C. (2005). *Educating the net generation – The student's perspective*. Retrieved August 18, 2011, from http://www.educause.edu/educatingthenetgen

Bibliography

Edwards, M. (2009). *Hard rain*. Retrieved August 24, 2011, from http://www.hardrainproject.com/

Watkins, M. (2014a). Towards an understanding of the social aspects of sustainability in product design: Teaching HE students in the UK and Ireland through reflection and peer learning. *The Journal of Design and Technology Education*, 19(1), 40–47.

Watkins, M. A. (2014b). *An investigation into effective methods for teaching social sustainability within product design in British and Irish Universities*. Thesis (PhD Doctorate), Loughborough University, UK.

Watkins, M. A. (2014c). Teaching the YouTube generation: Exploring the benefits of an interactive teaching approach in sustainable product design. In E. B. Sarrias, & L. M. S. Ruiz (Eds.), *Global Valencia* (pp. 501–506). Valencia: Escuela Técnica Superior de Ingeniería del Diseño Universidad Politécnica de Valencia

Watkins, M. A. (2015). An audio visual approach to the teaching of the social aspects of sustainable product design. *Formakademisk*, *8*(1).

Chapter 4
A Case Study on Problem-Solving Based on a Design Process in a Middle School Invention Class

Jae-Young Yu

Abstract The purpose of this research is to investigate the process of problem-solving through the design process in an invention class and then to analyze the effects of problem-solving and also to confirm the design process effect on learning achievement.

The subjects of this research consisted of four middle school students in two teams in a class in the gifted program. They met with the researcher as a team and completed assignments over 5 months. In order to analyze the behavior of the students, document analysis, interviews, and observations were conducted during the implementation of the class activities. This study investigated how students execute a design process that resulted from their problem-solving in the course of solving the conditions of the problems, through the process cycle of the performance, and how the upper domain and lower domain of the process influence each other.

The Questions I Asked and Why I Think They Are Important

A series of activities, defined as a design process, encompass creating ideas through brainstorming to solve problems, visualizing the created ideas into images with definite forms (sketch), showing the visualized images in isometric view and recording the measurements (conceptual drawing), proposing the list and number of the parts needed for the actual production (part drawing), and going through feedback processes while producing. This study aims to confirm whether problem-solving through design processes brings about a change in learning ability, as exhibited by students' experimental design process models.

J.-Y. Yu (✉)
Dae Ho Middle School, Gyeonggi-do, South Korea
e-mail: 3pmx018@hanmail.net

PJ. Williams, D. Barlex (eds.), *Contemporary Research in Technology Education*, Contemporary Issues in Technology Education, DOI 10.1007/978-981-10-2819-9_4

I observed how the design process changed as students participated in a problem-solving activity of producing a catapult. The assignment for the students was to make a catapult that meets two conditions: it should be able to change its angle and move a 10 g projectile more than 3 m. This assignment was an open format that could encourage divergent thinking from the students. For students to make a catapult which successfully meets those conditions, abilities to design and make it, as well as scientific knowledge, are simultaneously required. That is because the conditions are presented as a "nonstructured" form, demanding divergent thinking.

Also, we opened the computer lab for the students to conduct any preliminary research on catapult making. In the computer lab, students chose their own catapult type, which they thought was appropriate for the assignment.

After meetings over the period of 5 months with the students, I developed the interview questions based on my observations. Analyzing the student material (brainstorming notes, sketches, concept drawings, part drawings), I found some unclear designs and explanations and some content that was too abstract. Therefore, I initiated one-on-one meetings with each student to question them about any ambiguities in their material. The questions I asked them were in the four categories of their material development related to brainstorming, sketching, and concept and part drawings.

How I Tried to Answer the Questions

While they were being interviewed, using open-ended questions, the students' reactions were examined, and the data collected. Questions were restructured for them to be able to speculate on their experiences in the process of solving the problem.

While interviewing the students both in and out of the classroom, students asked a variety of questions. For example:

What would the process look like in order to produce a catapult that satisfies all the conditions of the problem?

How should Brainstorming, Sketching, Concept Drawing, and Part Drawing be expressed?

How would tools be used for actual production?

For each question, students were given specific examples that reflected a similar situation, instead of directly being told the answer. For instance, when students asked how to control the angle of the catapult, they were shown a concrete object to help them understand the meaning of adjusting angles and additional precautions when handling such matters.

Students were guided to respect each other's opinions, especially through the brainstorming process that was conducted in groups of two, so that they would share the same purpose of problem-solving throughout the project.

When students struggled coming up with a proper sketch, concept drawing, and part drawing, they were given a lucid explanation in detailed steps with the following questions:

How has humanity come to engage in drawing activities?
What are some advantages of pictures?
What are the differences between writing and drawing?

These questions were aimed at drawing students into self-directed learning, rather than supplying them with the answers. Through these processes, students were given an opportunity to acknowledge the transformation of the design process by working on sketches, concept drawings, and part drawings.

What I Found Out

So as to witness the change in problem-solving processes through the design process, the research was conducted with two groups of two (four in total) 8th grade volunteers in an honors class over a period of 5 months. Hutchinson and Karsnitz (1994) developed a design loop where the technology problem-solving process is

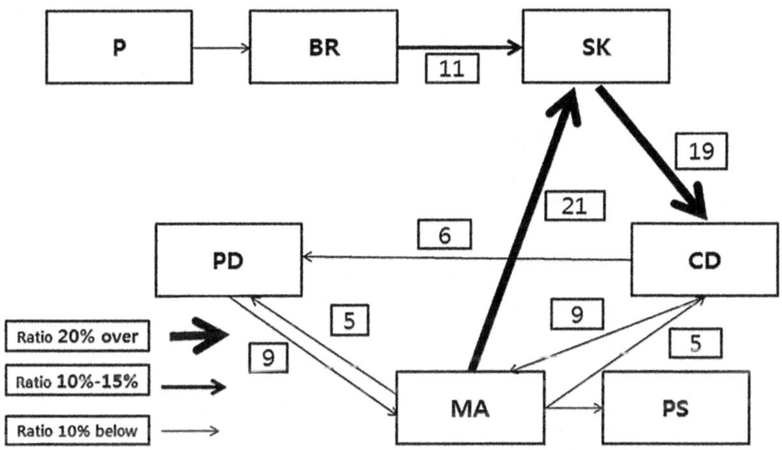

Fig. 4.1 Team A's execution process of the design process for problem-solving. *P* problem, *BR* brainstorming, *SK* sketch, *PD* part drawing, *CD* concept drawing *MA* making, *PS* problem-solving

nonsequentially interactive. They argued that the design could be repeated over and over again as needed. Therefore, quantitative and qualitative research is concurrently carried out to examine the more specific process of the problem-solving. The results are as follows:

1. *How did students execute the process that resulted from their problem-solving?*

Figure 4.1 depicts the analysis of team A's design process throughout the problem-solving activity, and it includes the number of repetitions of particular work done.

The highest numbers, 19 and 21 in the figure, indicate the frequency count of the same route that students took throughout the problem-solving activity. For instance, the number 21 on the arrow between making (MA) and sketching (SK) means that the students progressed from making to sketching 21 times because they failed to meet the requirement for the proper making of the catapult.

Failure in this paper is related to the processes (doing a repetitive work) rather than a consequential failure (observing the final product). The process could be likened to that of a toddler's learning steps. For example, a baby generally falls down around 2000 times (failure) until he/she stands on their own two feet and walks alone (final product). However, each stumbling and falling down is not just a failure but a very meaningful action. Accordingly, the students' repetitive work seems to be a process of the failure.

In the problem-solving via design process, the students intuitively practiced the ideas they thought of while they used tools and changed the designs. Of course, rather insufficient solutions are often derived due to their limited experience, and they go through several "trial and error" stages. It is, however, the very repetitive "trials and errors" that enable them to have a concrete idea based on the experience.

These indications should help understand where the students struggled the most throughout the process.

When students first confronted a problem, it was observed that both teams were equally inclined to proceed in the order of brainstorming (BR)–sketching (SK)–concept drawing(CD)–part drawing(PD)–making(MA) toward the completion of the final product. This is evidence that the students follow a nonsequential process when solving a problem, rather than simply going through the motions step by step.

Table 4.1 shows the categorized daily record of their own design process transformation. An analysis of students' activity shows that the design process is sequential or nonsequential. The sequential processes are BR-SK, BR-SK-CD, SK-CD, and SK-CD-PD; and the nonsequential cases are BR-SK-CD-MA-CD, SK-CD-MA-CD, MA-SK, PD-MA-CD, and PD-MA-PD. Numbers in parentheses indicate the number of the same work repetition.

Table 4.1 Team A changes the design process

<Table 1> Team A's Change the Design Process

1	2··	4	6	7	8	9	10	11	12	13··	17	18	19	20	42··	23	24·	27	28	29	30	31
December 2011																						
																				●BR		
January 2012																						
	※ ◎		※▲ BR											◎ ★BR							※ (4)	
February 2012																						
		▲(4) BR	●★ ◎ BR				■ (2) BR				★B R			★BR ♥(4)				● (4) BR		■(3) ★(2) BR		
March 2012																						
													◉▦	BR ■		■ BR			●(3)BR ★(2)		■ BR	
April 2012																						
								■ BR														
May 2012																						
			■(3) BR		■ BR				■ BR					■(5) BR								

●: BR–SK ■: MA–SK ◉: BR–SK–CD ▦: BR–SK–CD–MA–CD ◎: SK–CD ★: SK–CD–MA–CD ♥: PD–MA–CD

※: SK–CD–PD ▲: PD–MA–PD, BR: Brainstorming, MA: Making, SK: Sketch, CD: Concept Drawing, PD: Part Drawing

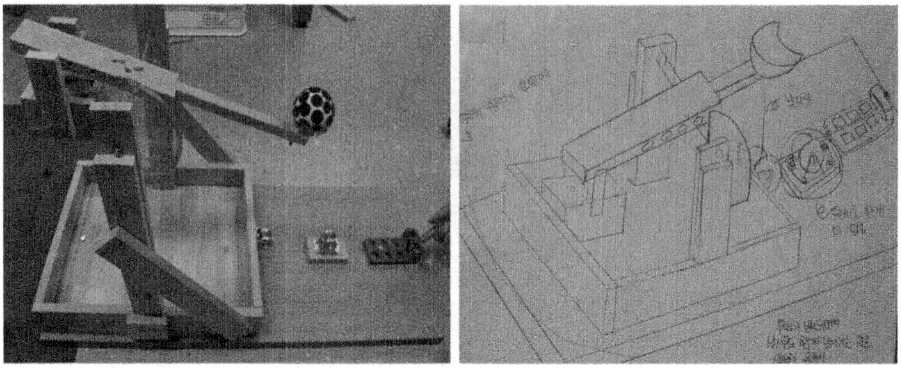

Fig. 4.2 The final completed catapults (Team A) and sketch

●: BR-SK indicates the design underwent some change in the middle of the sketching process.

●,★,◎: BR indicates that the students performed four tasks simultaneously on the same day.

★ means SK-CD-MA-CD. It is a concept drawing in sketch form, making in the concept drawing, and ongoing process of the drawing in making. As indicated in the table, students tend to repeatedly go back to the sketch again, after failing their making.

The analysis of the change in the design process shows that brainstorming did not just influence the early part but the whole design process throughout. Brainstorming was undoubtedly one of the most important factors in the problem-solving process, since out of 22 meetings, brainstorming was executed as many as 19 times (86 %). The number of meetings for team A and B differed because each team worked on a different type of catapult that required different scientific principles and production processes. Team A had the counterbalance catapult, whereas team B worked on the onager. Figure 4.2 shows the final product of team A.

2. *Do the upper (brainstorming, sketch, concept drawing, and part drawing) and the lower (making [observation, experiment, and measurement]) domain have an effect on each other?*

The upper domains of the design process focus on the activity performed before manufacturing the work. The lower domains revolve around checking the conditions of the problem as well as manufacturing the work. These subdivisions arose as a result of observations of a pilot study.

As a result of analyzing the students' problem-solving process during the catapult assignment, the lower domain (making [observation, experiment, and measurement]) of the design process influenced the upper domain (brainstorming, sketch, concept drawing, and part drawing) of the design process.

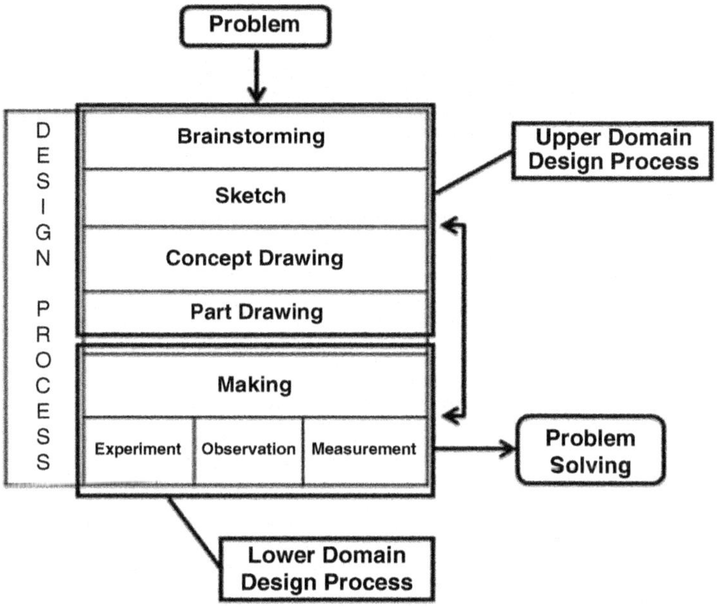

Fig. 4.3 The relationship between upper domain design process and lower domain design process

It was also verified that the upper domain (brainstorming, sketch, concept drawing, and part drawing) influenced the lower domain (making, observation, experiment, and measurement). This is shown in Fig. 4.3. The figure shows that each component in the design process individually works as well as organically influences each other. The fact that the lower domain has an effect on the upper domain means that students request redesigns (sketch, conceptual drawing, part drawing) in the upper domain when they cannot obtain their desired problem-solving results in the course of production. This process derives predictions of results by way of brainstorming. Also, the upper domain affects the lower domain as a result of the suggestion of necessary requirements for production, such as taking concrete shapes, taking arm measurements, and taking exact measurements of every part.

3. *The cycle is found in problem-solving process through the design process.*

When the problem was given to students, both teams A and B responded by taking the same path in the order of brainstorming (BR)–sketch (SK)–concept drawing (CD)–part drawing (PD)–making (MA), but this pattern did not happen sequentially for it went through the process of feedback, as Fig. 4.4 demonstrates. For instance, as Fig. 4.4 suggests, the sketch part has been continuously modified and enhanced throughout the problem-solving process. This implies that there have been numerous sketch changes during the process between the early and the final sketch. In fact team A has drawn ten sketches, while team B has four.

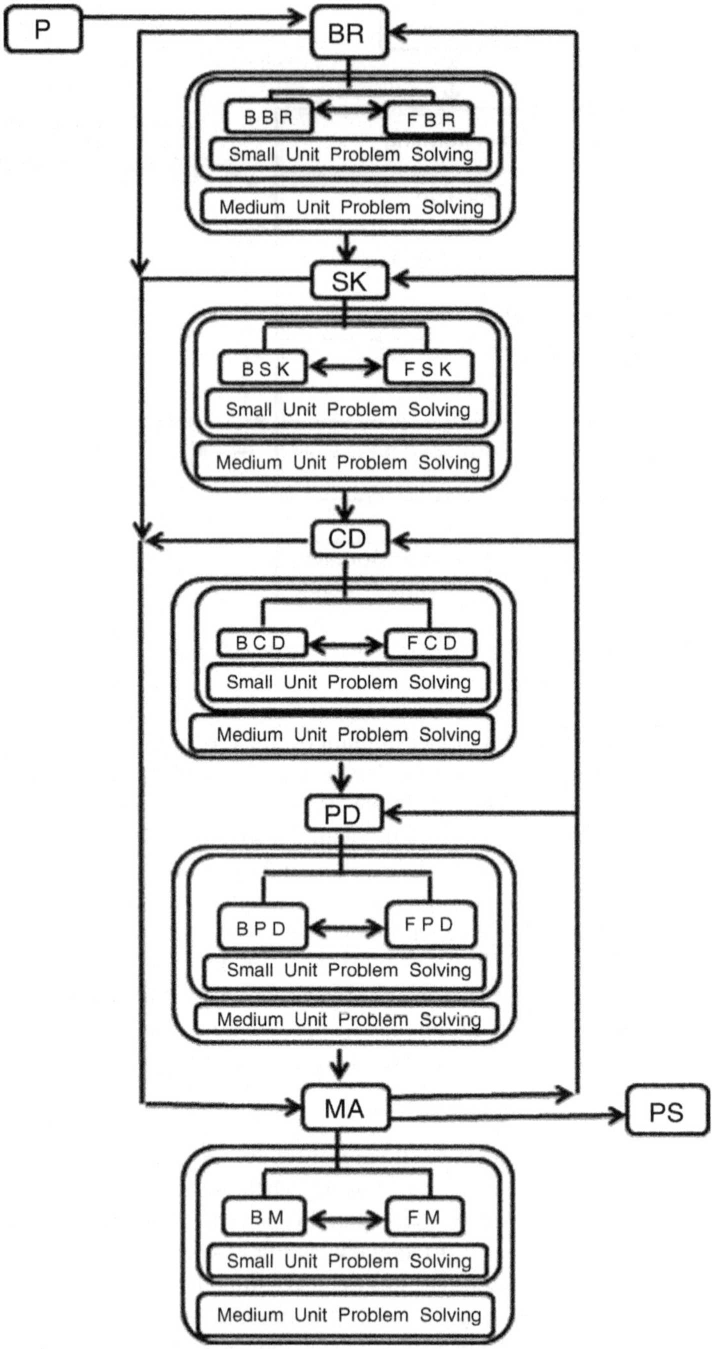

Fig. 4.4 Problem-solving model design process (*BBR* beginning brainstorming, *FBR* final brainstorming, *BSK* beginning sketch, *FSK* final sketch, *BCD* beginning concept drawing, *FCD* final concept drawing, *BPD* beginning part drawing, *FPD* final part drawing, *BM* beginning making, *FM* final making)

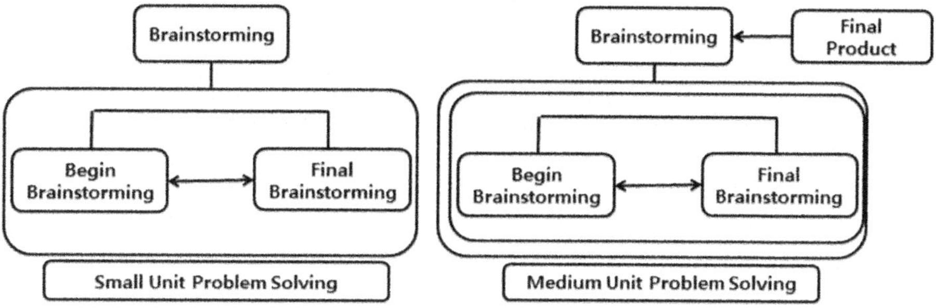

Fig. 4.5 Concept of a small-unit problem-solving and a medium one in brainstorming

Table 4.2 Developing a coding scheme

Design Process			Code	Definition
Design process upper domain	Brainstorming		BR	Brainstorming ideas regarding thinking, trouble, opinion, and conception to solve a problem
	Sketch		SK	Visualizing ideas into images with concrete forms
	Conceptual drawing		CD	Filling out the size measurement after visualized images turning into isoplane
	Part drawing		PD	Marking the type of parts and numbers on the drawing to make works
Design process lower domain		Making (experiment, observation, measurement)	MA	Checking whether it satisfies the conditions in problem after making works

Also, here are some examples of the brainstorming process as both a small-unit and middle-unit problem-solving agent. In order to make an appropriate catapult for the assignment, the students revised their ideas from the initial brainstorming stage through to the final brainstorming stage. Each brainstorming stage can be deemed as a small unit of the problem-solving process. In another perspective, the entire brainstorming process itself would be deemed as a middle-unit problem-solving agent for the whole assignment. Thus, brainstorming works as both a small-unit and middle-unit problem-solving agent in the designing process. Figure 4.5 elaborates on the meaning of a small-unit problem-solving and a medium unit in brainstorming.

And the same goes for sketches, concept drawing, part drawing, and making. Here, "the same" means that sketches, conceptual drawings, part drawings, and the production process all go through both small-unit problem-solving and medium-unit problem-solving to the same degree as in the brainstorming process. Therefore, it is confirmed that problem-solving through the design process is kept together (kept together means that a change of one design process brings about that of other design process) by the systematic mutual relations between all the components of the design process.

How this Might Be Used to Improve Teaching and Learning

What is important for the teachers who are studying design-based problem-solving processes is how to classify the variety of data output from students. Unless teachers have classification criteria, data produced from students will be wasted, disappearing meaninglessly. As there may be many unintended responses and outputs from the students, if teachers don't set appropriate criteria, support materials can be useless to the problem-solving research as it relates to the design process.

As seen in Table 4.2, a coding scheme was created to differentiate between the upper domain and lower domain of the design process, developing sub-elements of the domains. The code scheme is important as it helps confirm the way an idea develops, such as brainstorming, and how it is presented when the idea is encoded (pictured or recoded in a design daily record).

The research results imply that the advantage of the brainstorming process goes beyond thinking and conception (including a strong will to solve the problems and passion) which are not only simply presented as an idea for problem-solving but also a vibrant analysis of such problems can be included as follows:

How can the catapult be completed to meet the requirements of the problem?
What are the reasons for not being able to complete the requirements of the problem?
How can the problems, occurring in the process, be solved?

In addition, the time when brainstorming occurs is more frequent in the early stages where a problem is identified and the resolution of it is sought. According to the results of analyzing the contribution of brainstorming to problem-solving, it is confirmed that the brainstorming process plays a significant role in the problem-solving process.

Also, sketching, one of the upper domain skills in the design process, is a positively necessary stage for making the catapult, satisfying the requirements of the problem, and helping determine the requirements for catapult making. Additionally, sketching influences the making and completing of the catapult which satisfy the

Moving Distance Measurement of Projectile

1. weight of counterbalance basket is 500g; diameter of projectile 30gram, point of support.
1) in the case of weight of counterbalance basket is 500g; projectile 30gram, point of support

distance / angle	Moving Distance of Projectile (cm)			
	1st trial	2nd trial	3rd trial	average
90e				
100r				
110r				
120r				

Matrix

Fig. 4.6 Moving distance measurement, a down domain of design process, and matrix for confirming conditions

requirements of the problem, eventually helping understand scientific principles of what catapult itself has. Concept drawing of the upper domain in the design process is helpful in making by providing concrete dimensions such as arm length and stand width which are required in making the catapult, influencing pilot performance testing to solve the requirements of the problem.

The part drawing of the upper domain in the design process represents the correct value of each part required in making the catapult, which contributes to reducing error in making and eventually influencing early observation, experiment, and measurement.

The observation, experiment, and measurement components of making in the design process are playing an important role. For example, part of the matrix conducted in moving distance measurement, made by students, is provided in Fig. 4.6 with experiment conditions, variables, and measurement records required for filling out a matrix. In the course of these procedures, students can recognize that systematic design and activities should be involved to acquire the knowledge they want.

Changing students to become active learners, this process encourages students to get into a deeper level of activity beyond an experience of simply making objects. Through experiment, observation, and measurement, students can problem solve to meet the requirements of a problem beyond making a simple catapult. In addition, designing a matrix of writing and measuring experimental values in the making process can lead students to expand their thinking through experience.

In conclusion, the findings of this study have implications as follows: First, each of brainstorming, sketching, concept drawing, part drawing, and making in a course of problem-solving through design plays an important role and should be clearly recognized. Second, when students are in the middle of a problem-solving class based on their experience, teachers should know about systematic approaches to knowledge acquisition through experiment or practice. Finally, teachers should have rich experience about a variety of advantages and weaknesses of design and making.

What Might Be Investigated Further

Some advice can be offered for those teachers studying a problem-solving process through design as follows:

First, to achieve investigations into problem-solving processes, an appropriate teaching style and experience is needed. Through the progress achieved over a long period of time, a teacher's background knowledge and rich experience about problem-solving can inform an investigation into their pedagogy.

Second, technology teachers should have the capability to establish and operate effectively appropriate equipment setting, and the equipment for a class to engage in making as pre-operation safety of equipment is a variable that can make a great influence on not only students' safety but also effectiveness, appropriateness of class duration, and instruction.

Finally, among other important skills required as a researcher in a problem-solving process, a practiced eye to intuitively identify materials' fitness for process, carefully allotted class duration depending on the selection of materials, and observation skills to comprehend students' difficulty of the task will eventually contribute to improve teachers' capability of instruction through problem-solving class.

How Teachers Might Contribute to These Investigations

Some questions for teachers to consider include:

Can teachers recognize the differences between the problem-solving processes of ordinary and gifted students by examining their design processes?

Can teachers assert that gender, school environment, and background knowledge affect problem-solving processes through the design process?

References

Hutchinson, J., & Karsnitz, J. R. (1994). *Design and problem solving in technology*. New york: Delmar Publishers.

Chapter 5
CAD and Creativity at Key Stage 3

Deborah Winn

Abstract The use of CAD/CAM is now commonplace in education. Whilst some systems are relatively easy to master, the teaching and learning of 3D CAD can be problematic. Infrequent use means teachers may lack the confidence to teach it, and the complexity of the programs means students struggle to use it or be creative when using it. My research explores alternative methods of teaching 3D CAD from the perspective of both students and teachers.

The Questions I Asked and Why I Think They Are Important

Whilst it could be argued that there is little point in teaching students the more complex 3D CAD programs as easier to understand programs like 'Google SketchUp' also creates 3D models, I believe we should continue to teach the more complex programs from an early age as they more accurately reflect how things are designed and made. This prepares students more fully for a future workforce and as an informed consumer, which is important as stated by Sam Cox in Livingstone (2012). Cox says 'CAD design is used all over the place; you have to design something to build it. From that standpoint, it's important to give them the tools they need, and one of those tools is design software'. Cox also believes there are significant benefits to using CAD to teach math and science principles. 'CAD gives students a way to focus on working through a design process to understand what must be done to solve a problem'. Furthermore Cox (Cox Review of Creativity in Business 2005) recognises the need for creativity and innovation in design and business in order for the UK to be competitive in the world market. It therefore makes sense to provide students with the tools to be creative and use industry level technology from a young age to boost familiarity and confidence in this area.

It is true that technology changes rapidly and programs taught now are likely to be outdated by the time students reach the workplace; however the basic principles, features and what makes a model succeed or fail remain the same. The more

D. Winn (✉)
Neale-Wade Academy, March, Cambridgeshire, UK
e-mail: DWinn@neale-wade.org

PJ. Williams, D. Barlex (eds.), *Contemporary Research in Technology Education*, Contemporary Issues in Technology Education, DOI 10.1007/978-981-10-2819-9_5

complex programs mirror more accurately real-world applications and can be transferred more readily to other programs and functions such as photorealistic rendering and accurate orthographic and isometric drawings. I particularly enjoy watching students test model vehicles in a virtual wind tunnel or test the weights and structural integrity of their products, which would be less possible in the simpler programs. To some extent it solidifies the validity of the subject through cross-curricular activity and through truly extending students.

Initially a rapid enthusiasm to use the more complex programs after the introduction of CAD/CAM to the National Curriculum (Department for Education 1997) led to almost two thirds of schools in England being involved in the CAD/CAM initiative as reported by the Design and Technology Association in 2008. Since then, through discussion with other schools, its use seems to have decreased in favour of the more basic programs such as ‘Google SketchUp’ that are easier to use. I expect the reasons for this are that students and teachers appear to lose confidence easily when they struggle to make a model, and as students’ often only follow instructions to make identical models, they quickly lose interest. From a teaching perspective, CAD is only a portion of the curriculum and may be taught infrequently; therefore teachers often forget how to use the program, and they don’t have the time to learn its intricacies making it even harder to help students, which exacerbates the problem.

In essence I believe it is not what we are trying to teach but how it is taught that is the real issue and is a valuable area for research. The first question therefore was: can a more strategic based intervention improve the teaching and learning of 3D solid modelling CAD programs?

A considerable amount of research has been undertaken into the advantages and disadvantages of computer-supported collaborative learning, and it became clear at the start of the research that this approach could benefit students using these programs. The second question I tried to answer therefore was: can paired learning improve the teaching and learning of 3D solid modelling CAD?

My final concern when considering teaching methods was that students are generally asked to make identical products in a step-by-step fashion or by following one of the vast arrays of instructional video clips available on the Internet, which can be found via a basic search. This may give the teacher confidence as they would be able to anticipate possible problems that would make the model fail, but this is demotivating for students and teachers and does not follow the essential need to promote creativity in schools. My final question therefore was: can students be aided to achieve more creative outcomes when using 3D solid modelling CAD?

How I Tried to Answer the Questions

The research was carried out over four case studies. Three of these took place in the school in which I work and therefore adopted an action research approach. It is a large school with 300 students in each year group. All three of the studies were completed with 10–12-year-old students who were in their first year at the school, in order to reduce the impact of any preconceived ideas that they may have about CAD/CAM. Students were asked in the initial questionnaire if they had any experience of CAD/CAM, and at the time no one stated that they had, further reducing the possibility that they had preconceived ideas.

The first study, a pilot study, involved all students in the year group and attempted to ascertain how easy and how enjoyable the students expected using 3D solid modelling CAD to be. This is an important aspect of the research as Musta'amel et al. (2009) have indicated that 'perceptions that users have of CAD systems and their expertise can significantly influence their performance' (p. xx). It is an observation I have also made during lessons. The students all completed the questionnaire over 2 days as they were attending an induction event prior to enrolling at the school the following academic year. After a demonstration of the program, the students were given a questionnaire and were asked to score their expected enjoyment and ease of use out of 5. Spatial ability was tested via a simple spatial ability test given at the same time as the questionnaire. Finally, to test creative ability, students were asked to submit a design prior to the event which was placed in one of five categories as judged by a panel of three teachers based on how creative they believed the design to be.

The second study was undertaken after the same 254 students from the pilot study had enrolled at the school and had rotated through various design technology tasks lasting 6 weeks each. One of these taught PTC's Pro/Desktop 3D CAD program using the traditional 'follow my leader' step-by-step style of instruction. At the end of this course, the students repeated the questionnaire. Once the results were collated, students who had expected to enjoy using the program and did enjoy using it, students who expected not to enjoy using the program and didn't enjoy using it and students whose opinion had changed dramatically were invited to participate in semi-structured interviews in groups of six.

The third study took place in the following year at the induction day event, and the questionnaire was repeated with the new year group of age 11 students. This time once they had enrolled in the school, the same 3D CAD program was taught using the new method which was a simple game with rewards, played by pairs on a computer. Again the students repeated the questionnaire at the end of the course and the same groups of students were interviewed.

The last study was separated into two parts. To test the validity of the results, the same teaching method was then taught by another, less confident teacher in the same school to establish if teacher expertise and confidence influenced the outcome. Questionnaires were given at the beginning and the end of the course and the students were interviewed as before. In this study the teacher was also interviewed.

The final part of the study took place in a different school and was led by a teacher who was very vocal in that he did not like teaching CAD and who had very little confidence in teaching it. He had used traditional methods to teach Pro/Desktop but had found he was unable to assist students to create a successful model when they made mistakes. Again questionnaires were given out at the beginning and end of the course and the teacher was interviewed.

What I Found Out

The pilot study aimed to determine student attitudes before they used the CAD programs. Students were able to identify that there was a difference in complexity between the 2D and 3D programs. Most indicated on a Likert scale that they also expected to enjoy using the 3D program despite expecting to find it difficult to use, although to a lesser extent than the more simplistic 2D program. Students who used computers more for other activities are expected to find using the 3D program harder to use but also expected to enjoy using it more than those who didn't use the computer as much for other activities. It is difficult to determine from this study whether this was due to a more realistic understanding of the program or general enjoyment of computer-based activities as type of use was not established. In terms of gender there was little divide between the amount of time male and female students used the computer for, how difficult they expected using the program to be or how much they expected to enjoy using it. This is a significant change from previous studies in which female students appeared to be reluctant to become involved with computer-based technologies and is in contrast to studies by Cooper (2006) and Carter (2010) who concluded that the male/female divide still remains an issue because stereotyping exists. These studies suggest that gender issues are more task related as the girls still often show an 'I can, but I don't want to' attitude (Siann 1997, p. 120) when working with computers. This result was not taken at face value and gender difference was still considered in the later studies.

Sixty-eight percent of students didn't feel that making a product at the end of using the program would affect their enjoyment of using it; however this is an area worthy of more study as the reality may contradict this finding. Spatial ability was measured using a basic spatial ability test during the pilot study; however spatial ability did not appear to have any relevance against expected enjoyment or difficulty when using the program. The test was short due to time constraints; a more detailed test may have differentiated spatial ability more fully. Finally, creative ability was tested through submission of a hand-drawn design prior to the event which was categorised as one of five levels of creativity by a panel of teachers. The creative ability shown was surprisingly low considering the emphasis placed on creativity in the curriculum in England.

The second study compared the results of the pilot study student attitudes to using 3D CAD after they had completed a period of study using the traditional lock step teaching method. Those whose attitudes had changed significantly or had always been at the extreme ends of the scale were interviewed. Interestingly, the resulting answers were similar from all groups of students regardless of whether the students had enjoyed using the program or not, and often the answers given linked to known methods of encouraging creativity. Generally, students liked the idea of problem solving, and being set an achievable challenge, some suggested the possibility of levels that the students could work through. Many students wanted to work at their own pace as they felt they either had to wait for students or they felt rushed. Instructional video clips were suggested as they could suit most learning styles

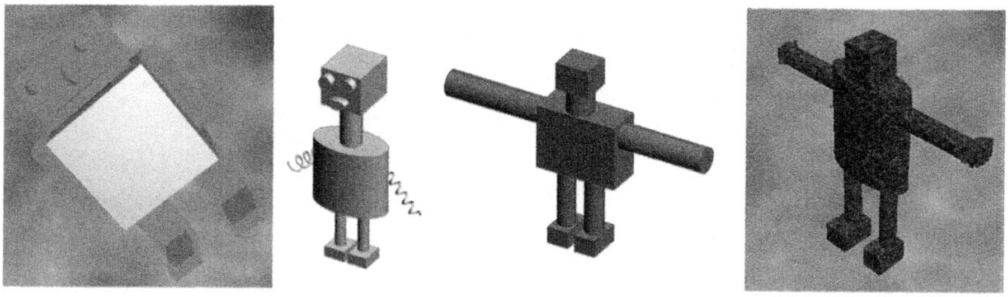

Fig. 5.1 Some of the most successful models from study 2

through the inclusion of audio, pictorial and written instructions. Students could also pause, play or rewind the clip to suit their own pace. Students also liked the idea of working in pairs as they believed they could discuss and work through problems more easily and that a partner would improve their confidence. Nearly all students were able to identify what personal creativity was but were unable to identify it in their own work, and little of their work was considered creative. One criticism however was that they had been asked to design a robot as it could be made using simple blocks and adapted with other features depending on ability, but it is possible that the stereotypical robot is too ingrained for students to think creatively in this area (Fig. 5.1).

From these interviews, the students stated what would help and motivate them, and an alternative method of teaching was created involving a scenario and levels the students could work through. In contrast to the teaching method in study 2, points would be awarded not only for getting it right but for a creative outcome and solving problems. The scenario and points were presented to the students as a simple game. Embedded in the game were videos that the students could work through at their own pace in pairs and a help section to aid the students to solve their own problems rather than wait for help.

Study 3 involved students working through the game suggested in the interviews by the students in pairs with the use of video clips; the same questionnaire was given before and after experiencing the teaching method as had been completed by the students in the pilot study and study 2. Interviews also took place after completing the course in the same way as they had in study 2. A far higher number of students expected to enjoy using the program and then did actually enjoy using it. This suggests that attitudes to the program have a vital role to play when teaching complex programs. The game aspect of the teaching method appeared to have given some students the confidence to try new things and had liked the challenge. After working through a variety of challenges, the students were asked to make a castle. Outcomes were vastly improved with a higher number of students producing a product, and most of these were considered creative by a panel of judges (Fig. 5.2).

Discussion between students throughout the lessons was encouraging. Students were using specific terminology, working through problems together and talking through possible creative options. At this point some of the pairs started working as individuals, choosing to make their own castle, although often still discussed their

Fig. 5.2 Some of the more successful models following study 3

Fig. 5.3 Some of the castles created in study 4

castle with their partner. Not all of the pairings worked however. Some students were observed by the teacher to let their partner do a majority of the work. However, if the discussion between the two students involved the program and the work I was not concerned by this, but pairings which did not were considered to be unsuccessful in terms of the paired learning aspect of the teaching method. Some pairs could not work together and continuously argued; others had problems if their partner was away.

The final study, study 4, involved two different teachers, in different schools, delivering the teaching method, and this is reported on from the teacher's perspective. Student outcome and attitude were reported on through these teacher's reports and the same questionnaires used in previous studies. Both of the teachers in these studies were less confident in teaching CAD, and one was very vocal in his dislike for teaching the program. Poor instructor confidence has been shown to be a negative factor for student confidence in previous research (Pektas and Erkip 2006); therefore this study is of particular interest. The classes of both teachers were able to complete the tasks set to a good standard and also showed a good level of creativity, which had pleased both teachers (Fig. 5.3).

The students had found the help sheets to be very useful, because rather than admitting they didn't know how to solve the problem could consult the help sheet. This is more beneficial to the student as working through the problem is likely to give the student a better understanding of the process than being told how to fix it. The first teacher reported that the students didn't appear to be as worried about getting things wrong as they had previously. However she found the leap from the structured task to the open brief too much for some students. This was similar to the experience of the second teacher who reported students had really enjoyed the more structured tasks but had not enjoyed the final task. Both teachers at times reverted to their known teaching methods. The first teacher included some step-by-step instruction with small groups of students, and the other teacher often used a whiteboard to demonstrate the task. The second teacher also wasn't confident enough to allow students to explore the program fully. This was not seen as a problem as the teachers largely followed the program of study as they had been instructed, and the additional resources gave the teacher and some students more confidence.

Paired learning had a mixed response with the teachers. The first stated she would only allow some, not all, students to work in pairs in the future as some of the pairings hadn't worked well. In contrast to previous literature reporting concerns regarding female students and computers, this teacher stated that the girls had worked better throughout the tasks and had achieved more favourable outcomes. It is possible that a female teacher, the style of the task and the paired learning helped to produce this result. The second teacher found that the paired learning was a positive influence and said it provided 'a far more valuable learning experience' and planned to adopt the approach when teaching other software. The most positive outcome of this study was that this teacher who had a dislike of teaching this 3D program had come to look forward to the lessons. He believed the resource removed the need for teachers to be 'experts' on the program and could remain updated on the programs if the software companies released this teaching method and released updates of it with any new software.

The main intention of this research was to investigate whether students could be taught to use complex 3D CAD programs more creatively in the early stages of learning the program by adopting an alternative teaching method in the same time period. The short answer to this question is 'yes' as many more students in the later studies achieved better and more creative outcomes; however there are several aspects to consider within the approach taken.

Student and teacher attitude rather than ability played a vital role in ensuring a successful outcome. Regardless of the teaching method used, low-ability students on several occasions produced more creative CAD models and were often more keen to 'give it a go' than some higher-ability students. The reasons for this may be worthy of further study and are suggested as such in the 'what might be investigated further' section. Those that expected to enjoy and do well at the task frequently did, and the attitude often did not change from before starting the task. I would have expected that the option to make a model of their own choosing, paired learning and the game aspect of the teaching method would have produced a more positive attitude to the task on all students than it did. The teaching method appeared to suit girls as well as boys and proved successful in this aspect as did rewarding effort as well as outcome. Being able to work at their own pace suited all students; however some do need to be reminded that this is not an excuse not to work and will need targets set to ensure they push themselves. The method, a game with rewards, played by pairs on a computer also suited inexperienced teachers or those lacking in confidence; however there must be some leeway for teachers to include their own teaching methods.

How this Might Be Used to Improve Teaching and Learning

There are several aspects of this research that are of benefit to the teaching and learning of computer-based technologies generally. However the main product to emerge is a resource to aid the teaching and learning of PTC's 3D CAD software Pro/Desktop and Creo. Currently limited to these programs, it would not be difficult to extend the resource to include other similar programs.

Whilst there are many other teaching resources available on the Internet for 3D CAD programs, they mainly focus on step-by-step instructions allowing students and teachers to complete a model. What they don't do is show how to solve the problems that users will undoubtedly encounter, which could affect the students willingness to try creative designs or attempt the task at all. As demonstrated in the study, some teachers will avoid teaching the programs if they do not feel competent enough to help students when they need to.

The difference with the resource created through this research is that it makes complex 3D design more accessible. It removes the need for teachers to become experts, allows students to work at their own pace and rewards effort and creativity as well as outcome. As technology changes quickly and teachers generally do not have the time to learn the new programs, it reduces the pressure for teachers to keep up to date as the software changes on top of other commitments and helps to increase the teacher's confidence as well as the students. The resource can be adapted by the teacher to suit their needs and preferences by only using parts of the resource or by adding demonstrations. Most importantly students were able to produce better quality models which incorporated more features than they had in the early studies using the software. They exhibited a higher level of confidence and were able to discuss their work with a greater level of understanding.

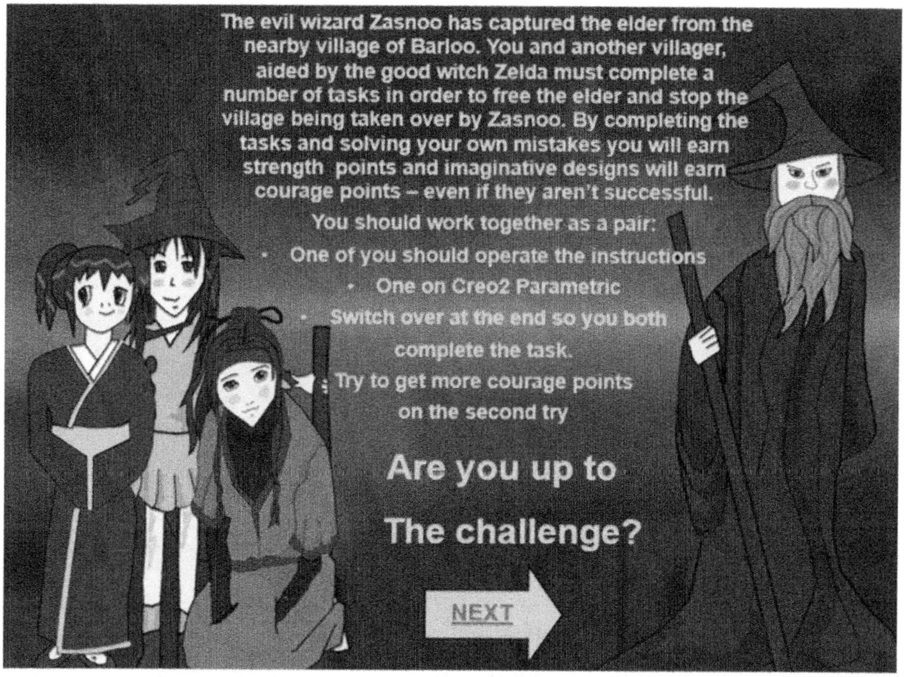

Whilst the resource appears to be little more than a PowerPoint with embedded 'show me how' videos, the workings behind the resource are what makes it useful. Although it could be made into a much more advanced game in the style of the type they may already play which records points and levels, this version still has many benefits. The scenario for the game is for students to battle a wizard and free a village elder by creating various items of varying complexity such as a key, coins, cup, etc.

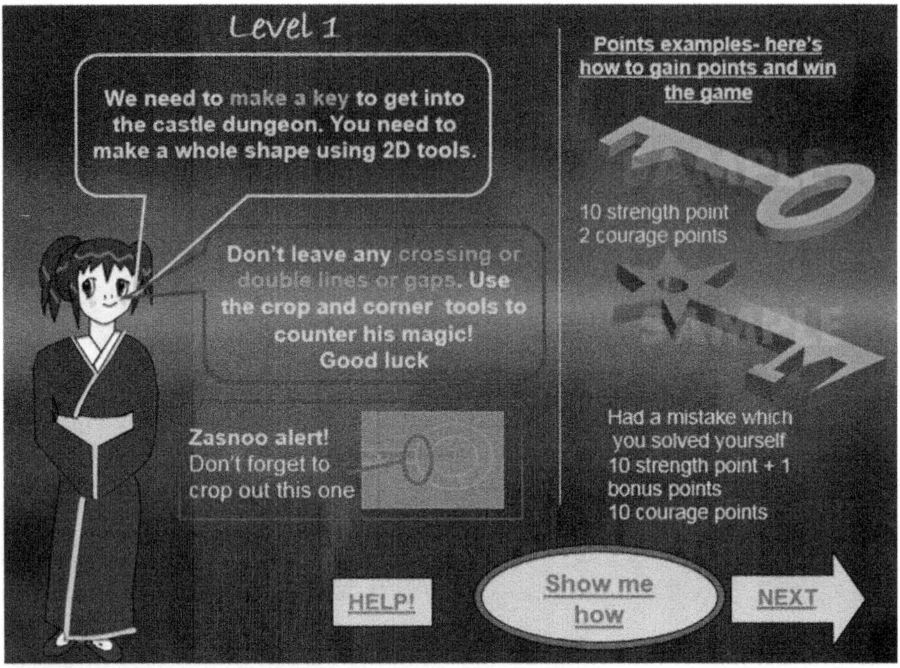

Points are awarded for what they achieve but more importantly for trying creative designs, even if they don't work, and for solving their own problems. This acknowledges that the problems are far more important than creating a perfect model but having no deeper understanding of how to create or adapt it.

To help the students (and teachers) find and solve the problems, they use a help booklet that is specifically written for each task.

Some students may never progress beyond the first couple of items. Others will work through all of the levels and be competent enough to create a complex castle with very little help. Most importantly all students will have a much deeper understanding of what the program is capable of and what might make a simple model fail whilst developing the confidence to try new things at their own pace.

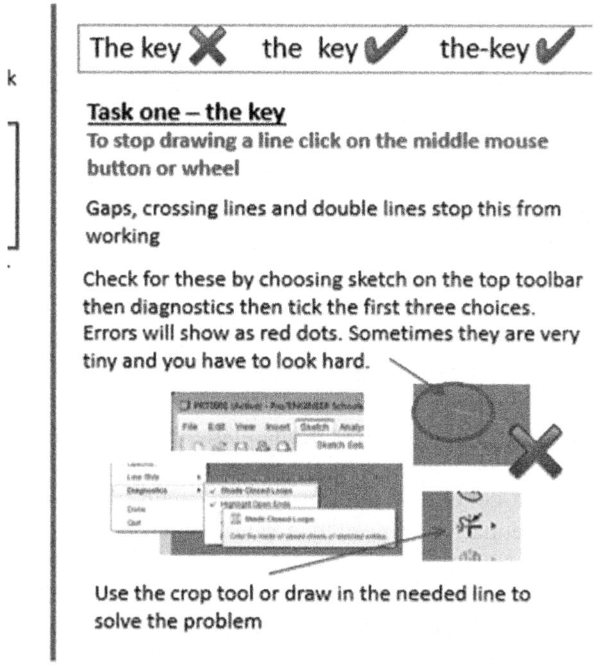

Without a resource like this, I believe many teachers and students will use the more simplistic programs such as 2D design and Google SketchUp. Whilst there are clearly benefits to using the more simplistic programs rather than nothing at all, I do not believe that they adequately represent how products are designed and made in industry, which is an important part of a students' education. They need to understand how products come into being in order to be an informed consumer able to make reasoned choices about the products they buy. Furthermore, new and expanding technologies such as 3D printers are becoming more common in homes, and an understanding of this technology and how to take advantage of its possibilities could be beneficial. Reporting on 'disruptive technologies' at the Computer Electronics Show 2014, Nunes and Downes (2014) reported that 'Analysts expect worldwide shipments of mass market 3D printers, already a billion dollar business, will nearly double from last year to almost 100,000 units in 2014 and double again in 2015'. They go on to report that a 3D printer can be bought for less than $500 and one made by Mcor Technologies uses normal printer paper as the medium for the product. This makes the printer very affordable to purchase and run potentially making it more likely people will want to experiment with the technology. The more complex 3D CAD programs also allow students to practise image rotation which assists to embed this vital technological skill.

The paired learning that took place in three of the studies in this research was largely collaborative learning as the students were responsible for theirs and others' learning although the students also worked together to create an outcome. The outcome although loosely defined so as to allow creativity ensured certain features were practised. In this resource the end goal is to experience what the program is capable of, become familiar with the technical language involved in this type of program and learn the common actions that are likely to make the model fail and how to avoid making them rather than complete a successful model. Providing students can see the point of what they are doing, a successful model is less important although obviously some success helps to motivate the students.

Paired learning was a vital component of the resource and could be useful throughout teaching computer-based technologies generally. It clearly benefitted most students, dialogue between students largely centred on the work and students worked together to solve problems given the appropriate amount of support; from discussion students seemed to understand what they were doing far more and generally gained confidence in their abilities.

Paired learning was also the most problematic element from the viewpoint of some of the teachers and students involved in the studies however. I believe the reason for this is that collaborative learning goes against how many teachers have learned to teach. Essentially you need to stop teaching and facilitate instead through the use of targeted questioning rather than through statements or instruction (Cohen 1994, Chiu 2004). Students can struggle to know how to work together, and others could use it as an opportunity not to participate fully in the task. To assist students this often involves teaching them how to work together whilst supporting rather than controlling the learning. This often presents a considerable risk for the classroom teacher but can be less risky given appropriate boundaries.

Quigly (2013) suggests the following for successful group work:

- Have clearly defined tasks, with sharp timings and with the appropriate tools organised.
- Have clearly defined group roles.
- Have clear ground rules for talking, listening and fair allocation of workload, etc.
- Target your support and interventions throughout the task, but make them interdependent of one another, not dependent upon you.
- Always be prepared to curtail group work if students don't follow your high expectations.

The activity presented to the students in this case followed these suggestions: it had very clearly defined tasks rising through levels of difficulty, and tools were presented both on the computer and in written format. The roles were also clearly defined in that one student controlled the video clips and helped their partner who controlled the CAD program. The students then switched roles, repeating the task, and were encouraged to be more creative on this attempt. It was important to emphasise to the students that they both receive points for both of the tasks thereby eliminating competition between the partners and making the outcome fairer. In the studies involving teachers less familiar with the program, they were only able to refer to the support materials and make reference to improving the creativity rather than instruct; however this did affect the confidence of the teacher in some instances. In what way precisely did it affect confidence?

The final suggestion is important for the success of this project, as times when the group work did not work well was when the student controlling the video clips did not pause and repeat the clips as needed and did not support their partner through the task. It was not sufficient to just play the clip but they have to instruct their partner throughout. It is imperative that these students follow the rules and are aware of the consequences.

This approach could be used for other Design and Technology tasks whether computer related or not given the appropriate resources and preparation. Possible noncomputer-related examples could be to provide creative modifications of other students' designs or provide an evaluation of another students' practical or design work which the first student would need to respond to.

Allowing students to be creative in this type of design work is a further risk for the classroom teacher. In England governing bodies in education demand that progress is made and that this is evidenced throughout the lesson. In this type of task where you are allowing students to complete work that may fail, it can be difficult to evidence their progress although leaps in their understanding may have been made by failing. To provide some evidence and to motivate students regardless of outcome in the game, 'strength points' were awarded for the success of the final model, and courage points were given for trying to be creative even if the model failed. It should be noted that the students should be able to identify verbally or in writing what caused the model to fail.

The use of reward systems when teaching is not new; however rewarding effort and problem solving rather than outcome is less common. Students need to be less goal orientated in order to take the risks necessary to learn complex programs and be creative in their designs. The focus needs to be on the process and controlled mistakes must be seen as part of this process rather than an element to be avoided. One of the teachers in study 4 reported that the students abandoned copying the videos to achieve a more creative possible outcome as soon as they realised they gained more marks for producing a model different from the one given.

Although process is more important than outcome, as stated earlier, some success in achieving a model is necessary to motivate the students; therefore it is important that the mistakes are controlled to some extent as the task has to be achievable and the students must recognise what the mistakes are and have an opportunity to solve the problems that arise in order not to be demotivated by the task. This was seen clearly in the game element of the resource where student's enjoyment and confidence increased during the more structured tasks but decreased in the open task where support was removed.

What Might Be Investigated Further?

Some surprising results in terms of motivation, attitude and outcome would benefit from further study. In particular why some low-ability students are able to become proficient at the program and produce highly creative outcomes easily when high-ability students sometimes seem either unwilling or unable to use the program. This is clearly the result of attitude and expectation and a more in-depth study researching this would be of value throughout the teaching of any complex software and possibly other tasks in which the student 'can't do it' but is in reality capable. An aspect that may affect student attitude is how much and for what purpose students use computer technology outside of school. It may be that the current trend for social networking sites and gaming, which often also involves dialogue with others, makes a difference to attitude.

This study indicated that there are many benefits to students working together when learning computer-based technologies; however further research into effective pairings and paired learning would be useful throughout all areas of teaching and learning. A final consideration for further study would be more in-depth research into any link between spatial ability and the ability to use complex 3D CAD programs. This study used a very short spatial ability test which appeared to indicate that spatial ability made little difference to the students' ability to use the program. However it seems reasonable to assume that spatial ability must have some kind of bearing on how well the student is able to visualise the 3D product mentally and on screen and be able to manipulate it effectively.

How Teachers Might Contribute to These Investigations

Design and technology lessons involve applying knowledge largely through doing and research in this area is no different. Most of the possible areas of research could be completed through observation and reporting on normal lessons in school. There are many benefits to this being completed by a practising teacher who would understand the dynamics of the classroom. However there is an argument to be made for having someone with fresh eyes or alternative skills observing the classes.

To an extent for existing teachers, the 'having a go' approach would be beneficial, and many good teachers already do this in their classes, as a form of action research. They try out different strategies, assessing their effects on students or groups of students, and then adapt the tasks or resources accordingly.

One area this approach could prove enlightening is in providing activities which could potentially improve the student's ability to mentally hold and rotate images and comparing their subsequent CAD ability to those that haven't had access to these resources then adapting further resources for all students.

In-depth case studies of high computer and spatial ability students who are particularly able would also be beneficial. Detailed questionnaires and Likert scales to determine their preferences and what makes the student confident in these tasks could be revealing and a good starting point for any research. In a similar way a detailed spatial ability test may provide further insight into why some students seem to become proficient with the program so much faster than others.

Acknowledgement I would like to express sincere gratitude to my supervisors during my PhD, Frank Banks and Steve Garner, in particular to Frank as I would not have been able to complete my studies without his insightful help and support, and also to Neale-Wade Academy, Richard Dunkerly from Ramsey Abbey school and Tim Brotherhood from PTC who assisted during the studies to various degrees. Finally, a thank you to friends and family who supported me through my PhD studies.

References

An electronic copy of the PhD Thesis can be found at this url: http://ethos.bl.uk/OrderDetails.do?uin=uk.bl.ethos.580671

Carter, J. (2010). What are the important gender-related issues in computing at present. *Computer Science Education, 20*(4), 261–263.

Chiu, M. M. (2004), Adapting teacher intervention to student needs during co-operative learning: How to improve student problem solving and time on task. *American Educational Research Journal, 26*(1).

Cohen, E.G. (1994). Restructuring the classroom: Conditions for productive small groups. *Review of Educational research, 64*, 1–35.

Cooper, J. (2006). The digital divide: The special case of gender. *Journal of Computer Assisted Learning, 22*(5), 320–334.

Cox, G. (2005). *Cox review of creativity in business: Building on the UK's strengths*. Retrieved April 10, 2008, from http://www.hm-treasury.gov.uk/cox_review_creativity_business.html

Cox, S., in Livingstone, H. (2012) http://www.cadalyst.com/cad/get-'em-while-they're-young

Department for Education. (1997). *Design and technology: The national curriculum for England*. London: HMSO.

Musta'amal, A. H., Norman, E. W. L., & Hodgson, T. (2009). Gathering empirical evidence concerning links between computer aided design (CAD) and creativity. *Design and Technology Education: An international Journal, 14*(2), 53–66.

Nunes, P, & Downes, L. (2014). http://www.forbes.com/sites/bigbangdisruption/2014/01/10/the-five-most-disruptive-innovations-at-ces-2014/. Accessed 11 Mar 2015.

Pektas, S. T., & Erkip, F. (2006). Attitudes of design students toward computer usage in design. *International Journal of Technology and Design Education, 16*(1), 70–95.

Quigley, A. (2013). *HuntingEnglish*. Retrieved March 11, 2015, from http://www.huntingenglish.com/2013/01/12/top-ten-group-work-strategies2014

Siann, G. (1997). We can, but we don't want to: Factors influencing women's participation in computing. In R. Lander & A. Adam (Eds.), *Women in computing*. Exeter: Intellect.

Chapter 6
Mixed Media Modelling of Technological Concepts in Electricity

Sarah Pule

Abstract Conceptual and perceptual constraints can result in the lack of transfer of knowledge. This research presents a variety of novel, custom-designed learning aids employed to rectify the lack of transfer of knowledge for simple resistor-capacitor circuits. The design of these learning aids is based on the concept of embodied cognition and mainly makes use of visual and kinaesthetic means to appeal to students who may have diverse learning styles.

The Questions I Asked and Why I Think They Are Important

The research aim of this study is to recognise the learning styles of engineering and technology students and propose pedagogical methods for the comprehension of technological concepts in electricity.

The nature of the field of electricity makes the subject very singular to teach since learners need to develop skills of representing the same circuit in several ways, such as schematic, prototype, mathematical or simulation. The development of these skills necessitates continuous shifts between conceptual representations of the actual circuit and may prove to be different for people with diverse learning styles. Novice students may develop biased conceptual representations of circuit behaviour, especially when the circuit is slightly changed in its connectivity structure. External representations influence the understanding of concepts. This research explores how external representations may be designed to help students with different learning styles achieve successful comprehension of resistor-capacitor series circuits. A list of research questions addressed by this study can be found in Table 6.1.

Although this research specifically deals with the behaviour of resistor-capacitor circuits within the domain of electricity, it is worth taking a broader perspective and reflect on the wider implications of similar questions, as these may apply to other topics within the domains of engineering and technology. The asking of analytical

S. Pule (✉)
University of Malta, Msida, Malta
e-mail: sarah.pule@um.edu.mt

PJ. Williams, D. Barlex (eds.), *Contemporary Research in Technology Education*, Contemporary Issues in Technology Education, DOI 10.1007/978-981-10-2819-9_6

Table 6.1 The research questions addressed in this study

1	2	3	4
	Objective: to enquire about learners' prior knowledge base about RC circuits		Objective: to determine if the teaching intervention has had any effect
	Pretest questions	Teaching intervention	Post-test
Analytic part	1. What verbal or diagrammatic representations are present in learners' schemata about the topic of simple RC circuits resulting from prior learning?		2. What is the influence of the novel teaching intervention in this study upon the verbal and diagrammatic representations present in learners' schemata about RC circuits?
	1.1 How are these linked?		2.1 Do the participants exhibit more evidence of intellectual growth, of scientific explanation and of expert-like behaviour within and beyond the possessed schemata?
	1.2 Which analogies do they include?		
	1.3 If the schemata include the circuit schematic, what is the preferred way for visualising the circuit schematic diagram? As a diagrammatic representation, could this be linked to the analogies included in the schemata?		
	1.4 How may variations in the topology of the circuit schematic influence the transfer of knowledge from topologically and conceptually similar circuits to other circuits which are conceptually similar but topologically unfamiliar?		
	1.5 Do graphical representations feature within the schemata? Which particular ones? Do participants develop visual-verbal paired associates between graphical representations and verbal technical jargon? Is this a help or a hindrance?		

	Objective: to enquire about learners' prior knowledge base about RC circuits		Objective: to determine if the teaching intervention has had any effect
	Pretest questions	Teaching intervention	Post-test
	1.6 Is there evidence of cognitive flexibility, scientific explanation and expert-like behaviour within and beyond the possessed schemata?		
Design part		3. How may learners be aided in constraining the problem search space about simple RC circuits?	
		3.1 How may the dynamic abstract concept of an electrical circuit be visualised qualitatively by using its circuit schematic as an external representation?	
		3.2 Do learners have preferences for the design of models?	

questions is important. It is only through such an approach that a clearer picture of student characteristics and any prior knowledge is revealed, including misconceptions and distorted thought processes. Asking pedagogical design questions is equally important because it makes a teacher reflect on methods of teaching and how these might be engineered for different purposes with different audiences. These are the two terms that are particularly pertinent to this research: (a) threshold concept and (b) habits of mind. The characteristics of a threshold concept are articulated by Meyer and Land (2003): a threshold concept is transformative and once understood, changes the way a learner views the discipline. The broad impact of this research lies in considering the behaviour of RC circuits as a valuable threshold concept for engineering and technology education as suggested by Peter Goodhew (Goodhew 2010) and his '*would be nice to know*' list for freshmen engineering students. The concept of engineering habits of mind features in Thinking like an Engineer (Lucas, Hanson, and Claxton 2014). Habits of mind are typical thinking behaviours that engineers employ frequently, for example, systems thinking and creative problem-solving amongst others. The habit of mind that is particularly relevant to this work is 'visualising'. Within a discussion about engineering habits of mind, the broad impact of this research lies within the pedagogical design aspect and how engineering habits of mind such as visualising and systems thinking may be developed for novices in the field.

How I Tried to Answer the Questions

A mixed method approach was used for the data gathering process. The main data collection method consisted of a structured interview. Within the interview, a test was used and observations were collected. The discussion conducted during the interview was led by the questions presented in the test. These were closed questions, requiring factual answers and not opinions, and concerned the topic of resistor-capacitor circuits. The objectives of the test questions can be listed as follows: (a) to test the familiarity of students with foundation concepts such as operation of SPST and SPDT switches, calculation of potential division and reading of voltages; (b) to envision positive or negative voltages for a node in a circuit; (c) to identify the resultant shape of graph when the circuit is made to go through a dynamic change, for example, switching over a SPDT switch, (d) to match the state of the circuit to a graph shape; and (e) to match the state of the circuit to a verbal statement.

Participants

This research involved three groups of participants: electrical engineering year I (EEY1; N=38), electrical engineering year II (EEY2; N=37) and design and technology postgraduate teachers (DTPGT; N=7). These engineering and D&T groups were selected because the research necessitated individuals who had been previously exposed to advanced knowledge, as defined by Spiro et al. (1988), on the topic of resistor-capacitor circuits.

The Structured Interview

The structured interview was conducted in two sessions, the pretest and the post-test, each of a nominal duration of 2 hours. During the pretest session, participants were asked to fill in the Felder-Silverman Index of Learning Styles questionnaire (Felder 1988), together with their demographic details. They were then given a sheet on which to draw a resistor-capacitor circuit using given symbols and any analogy of a capacitor with which they were familiar. Participants then started working on the test questions.

The post-test phase involved the viewing of a 15 min video specifically designed to complement other teaching resources. The teaching intervention was then initiated. This lasted approximately 20 min for each participant. Concepts in the video were reinforced through the use of the tangible, qualitative resources. The main aim was to present *the relative-dynamic two-plate model analogy* for the behaviour of a resistor-capacitor circuit in order to get the participant to engage in a discussion about the behaviour of all circuits, using this analogy. The design concept underlying all tangible resources was the visualisation of the voltage potential for salient nodes in the circuit, with special focus on both capacitor plates.

Participants then attempted the questions in the post-test. These questions were identical to those presented in the pretest. Throughout the post-test, participants were encouraged to interact with all the teaching resources at will. A taxonomy of resources utilised for the teaching intervention is represented in Fig. 6.1.

The 'Relative Dynamic Two-Plate' Analogical Model of an Electrical Capacitor: The Concept Underlying All Teaching Resources

The *relative-dynamic two-plate model* consists of representing the capacitor by its standard schematic symbol as shown in Fig. 6.2.

The variety of states which this analogy can accommodate are illustrated in Fig. 6.3. The voltage difference read across the plates represents the voltage across the capacitor in a specific state. The vertical distance in between the plates represents the voltage difference between the plates, which indirectly infers 'volumetric quantity' contained in the capacitor.

This analogy is different from the common water tank analogy because it is abstract and not tied to a real-life mechanical system. Although the function of containment can be perceptually inferred, the relative-dynamic two-plate model is more

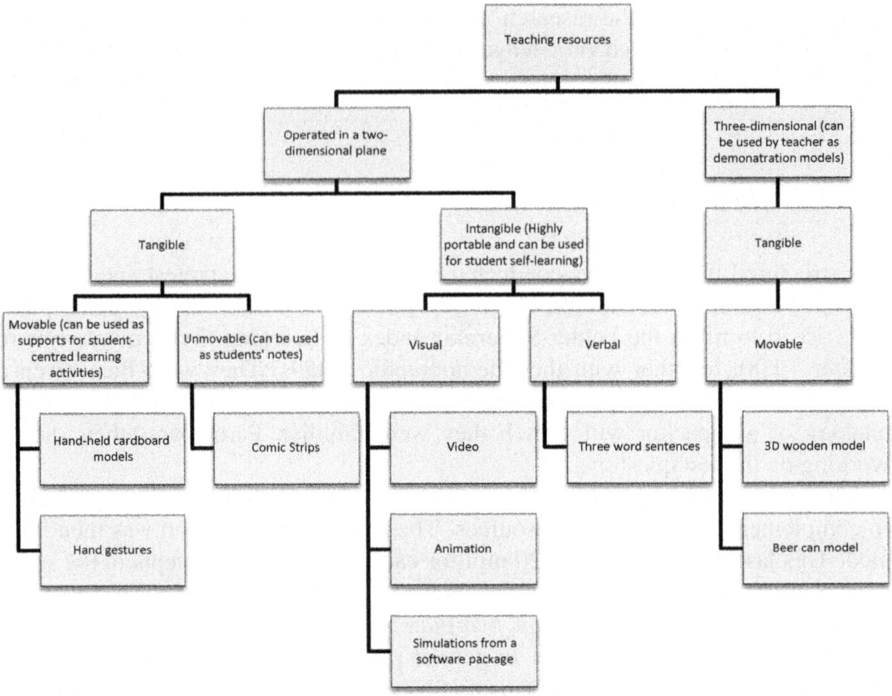

Fig. 6.1 A classification of the teaching resources

Fig. 6.2 Diagrammatic representation of the capacitor within the *relative-dynamic two-plate* analogical model

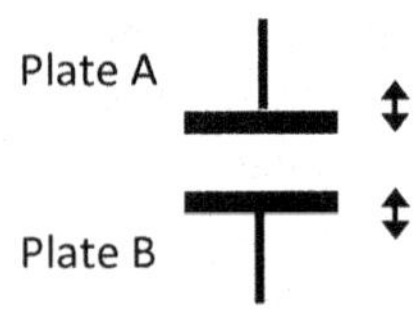

suitable in describing the coupling effect governing capacitor plates when they are subjected to transient signals. The resources designed and implemented in this research all adopt the analogy presented by the *relative-dynamic two-plate model*. Photographs of the resources utilised are provided in Table 6.2.

What I Found Out

Participants' Learning Styles

Participants' learning styles were measured using the Felder-Silverman Index of Learning Styles (FSILS) questionnaire (Felder 1988). The bar charts of Fig. 6.4 show the outcome of each of the four dimensions of the (FSILS) for all participants.

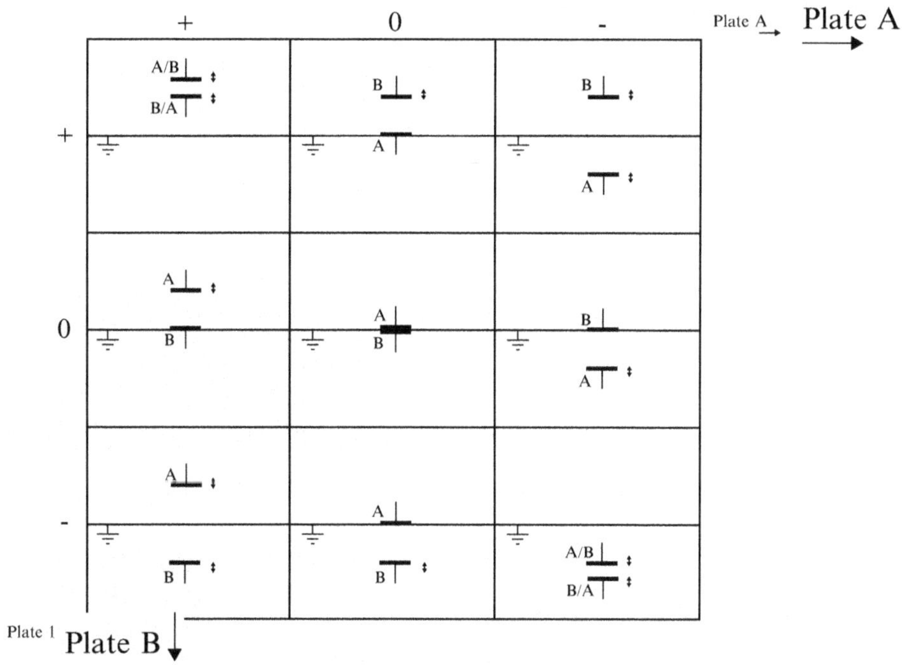

Fig. 6.3 Possible states of the capacitor plates within the *relative-dynamic two-plate model*

The distribution indicates that this study's participants were fairly balanced on the active-reflective dimension; the majority is of the sensing type; a staggering 97.32 % are visual learners and, overall, most participants are sequential learners.

The overall learning style scenario has shown that the majority of learners are, in order, active, sensing, visual and sequential. These findings concur with Felder's observations on engineering students (Felder 1988).

Most of the teaching resources were intended to cater for the active, sensing, visual type of learner since they took advantage of participants' visual perception and kinaesthetic modality of receiving information. This was achieved by designing the resources as visual, tangible, movable artefacts, whose salient features could easily be recognised by their resemblance to the familiar, standard electrical symbol schematics used for the components represented, i.e. a resistor and a capacitor.

Table 6.2 Photographs showing the teaching resources

	Resource description	Image of resource
A.	The hand-held cardboard model	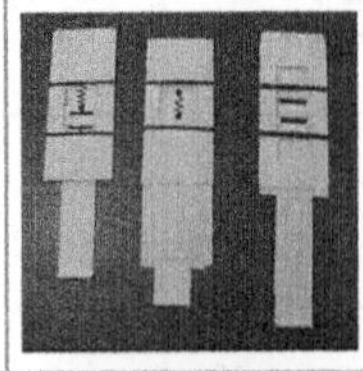
B.	The animation where circuit components are superimposed on the graph (seen on video)	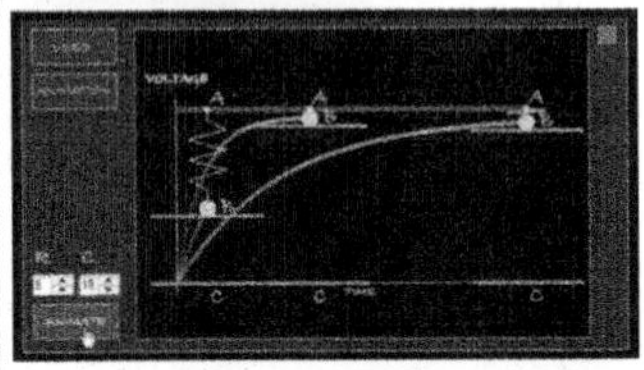
C.	The simulation from a software package	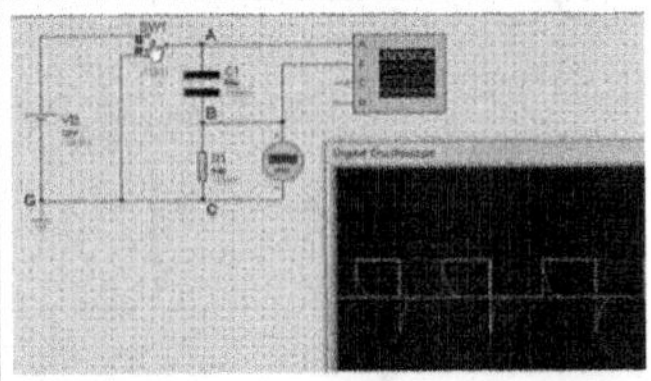
D.	The beer can model	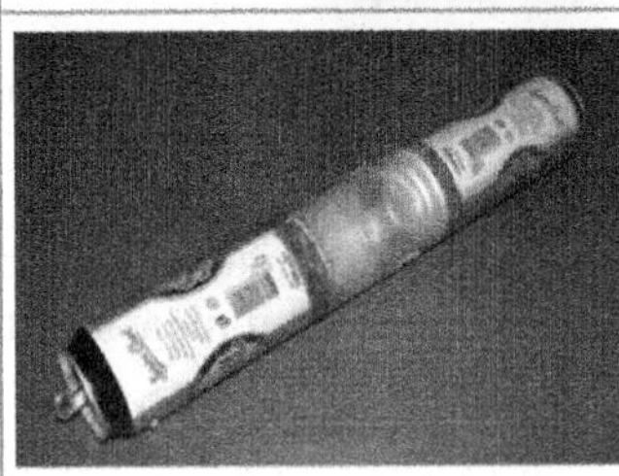
E.	The comic strips	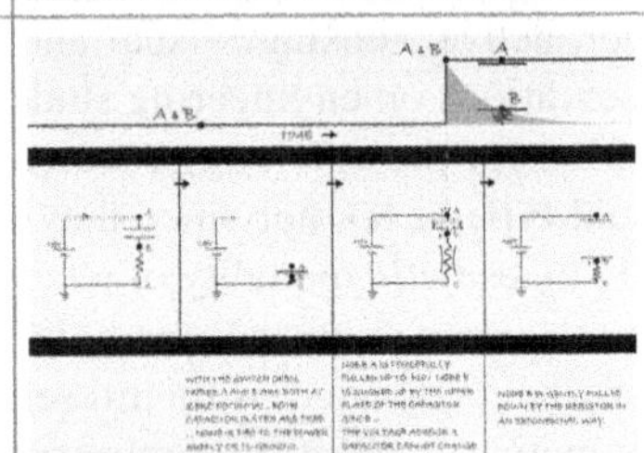
F.	The 3D wooden model	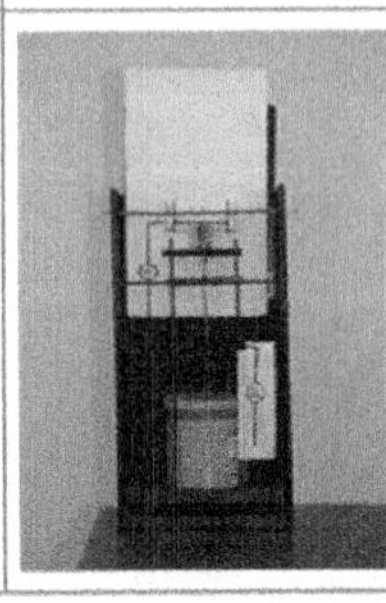

(continued)

Table 6.2 (continued)

	Resource description	Image of resource
G.	The three-word sentences given as guiding steps	*1. Observe the circuit diagram well ... check which nodes are strongly attached (maybe to the terminals of the supply) ... and which nodes can float*
		2. Determine the initial state of the capacitor ... Charged or Discharged
		*3. Apply the GOLDEN RULE: ... 'the voltage across a capacitor **CANNOT** change very quickly'*

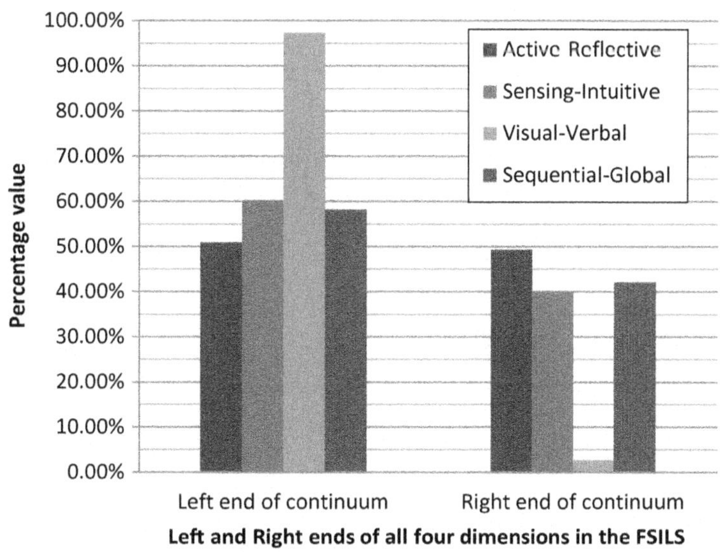

Fig. 6.4 Collective percentages of each of the four dimensions of the FSILS

Relationships Between Common Analogies Used for RC Circuits, Electrical Circuit Schematics, Technical Jargon and Corresponding Voltage-Time Graphs

The data analysis process was approached through the hypotheses listed in Table 6.3 (Figs. 6.5 and 6.6).

Figure 6.7 represents the participants' conceptual framework on the topic of simple resistor-capacitor circuits at the pretest stage. Participants' preferred analogy for an RC circuit was that of the *stand-alone water tank*, or *bucket*. Participants had

Table 6.3 Researcher's hypotheses

Hypothesis 1	Most participants would tend to have preconceived ideas regarding the shapes of graphs associated with the output of particular circuits. More specifically, it was hypothesised that participants would tend to always associate the graph shape of Fig. 6.5 with a charging capacitor and the graph shape of Fig. 6.6 with a discharging capacitor, irrespective of the configuration of the circuit
Hypothesis 2	As a consequence of hypothesis 1, participants would experience conflict when they encountered circuit situations which challenged their preconceived ideas about circuit behaviour and corresponding output graph shape
Hypothesis 3	Affordances are visual clues to the function of objects. The affordances of the circuit schematic and the corresponding graphical representation would influence the participants' responses as conjectured in hypothesis 1. The affordance of the circuit schematic relates to the presence or absence of a voltage source in the loop formed by the circuit. The affordance of the graphical representation relates to its piecewise slope
Hypothesis 4	Most participants would not be aware that, for some particular circuit configurations given, the output voltage could rise above or fall below the voltage range of the ideal voltage source given in the problem

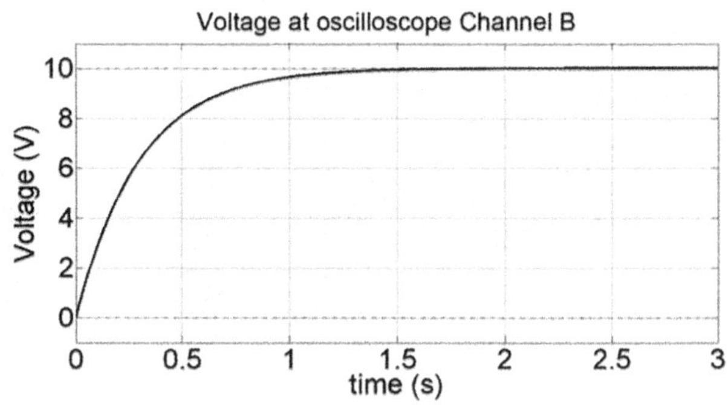

Fig. 6.5 Exponential graph associated with the charging of a capacitor

a preferred pattern of how to draw the circuit schematic diagram, and the reason for this may stem from the analogical structure mappings with the *water tank* analogy (refer to Section 4.6.2.1.2 page 162).

Pule' (2012) shows under which circumstances a student tends to use graphs A, B in Fig. 6.7 as conceptual pegs to the words '*charging*' and '*discharging*', even when the functional information gathered from an intermediary source such as the circuit schematic is in opposition to their visual-verbal associations.

Particular attributes of the diagrammatic displays of the graphical and schematic representations of the electrical circuits may be working together with the use of technical jargon in relation to the success or, otherwise, of the transfer of knowledge. Apart from the influence of analogy, the following four variables act together when participants attempt to transfer their knowledge about simple RC circuits to unfamiliar situations: (a) the words *charging* and *discharging*, (b) the relative positions of the symbols of resistor and capacitor in the circuit schematic, (c) the piece-

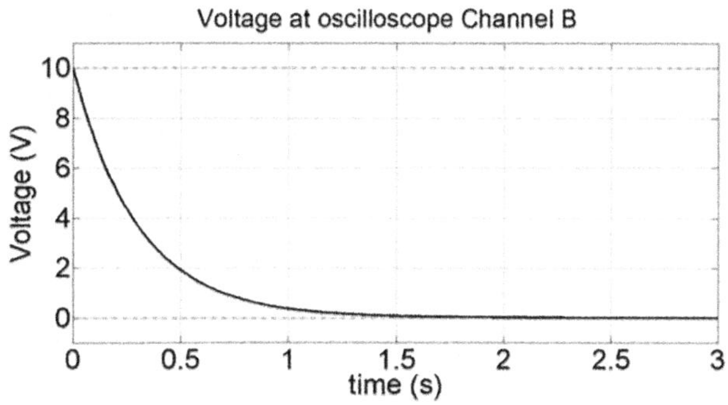

Fig. 6.6 Exponential graph associated with the discharging of a capacitor

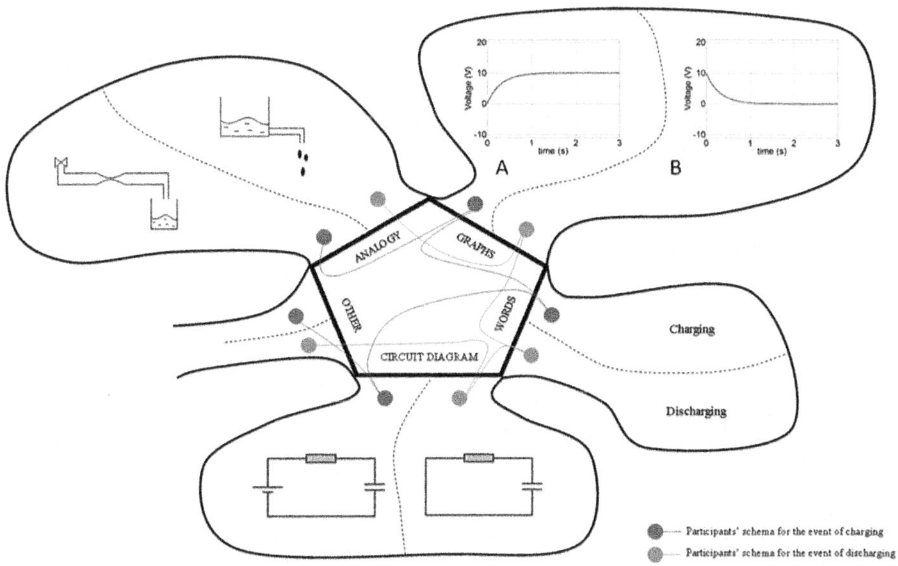

Fig. 6.7 Participants' schemata prior to the pretest

wise slope of the exponential graphs and (d) the presence or absence of a voltage source within a loop.

Ranking Students' Preferences for the Novel Resources Used During the Teaching Intervention

Figure 6.8 leaves no doubt as to the participants' self-rating of their performance after experiencing the teaching intervention and an opportunity to tackle the same questions a second time. Clearly, they felt that the teaching intervention positively

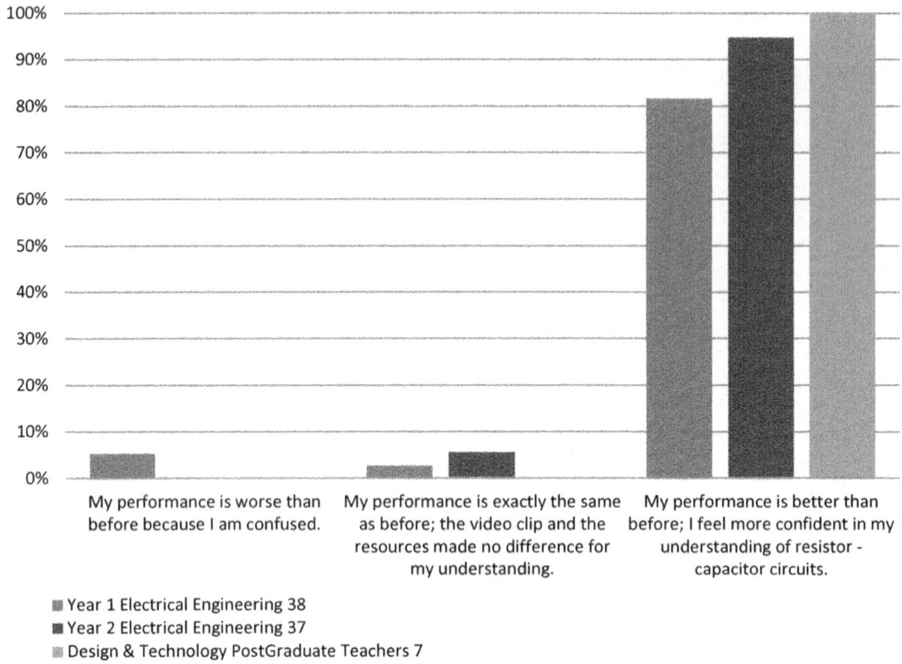

Fig. 6.8 Participants' self-rating of performance after post-test

influenced their performance and made them feel more confident about their understanding of RC circuits.

Figures 6.9 and 6.10 show that the video was perceived as useful and helped raise participants' confidence levels. Nevertheless, further analysis of the ranking which participants gave to all resources showed that the tangible models were preferred over the intangible ones as indicated in Table 6.4. Participants' confidence started increasing during the teaching intervention after watching the video, but was confirmed highest only when participants had direct experience with the more tangible resources of the set provided.

For all three groups, the hand-held cardboard models and the *3D wooden model* ranked first in preference, while the simulations from a typical electronics software package and the verbal statements were ranked last in preference.

This result confirms that participants had a strong preference for visual, tangible models rather than intangible or more abstract ones. This outcome can be explained by the results obtained for students' learning styles as measured by the Felder-Silverman Index of Learning Styles. Participants' choice of visual, tangible media supports the outcome which states that their learning styles are active, sensing and visual.

Participants found that using the resources was not complicated and time-consuming, once they were helped in their interpretation and operation. Although simple in nature, the resources still showed circuit behaviour in a certain amount of

Fig. 6.9 Participants' confidence level after watching video but prior to post-test questions

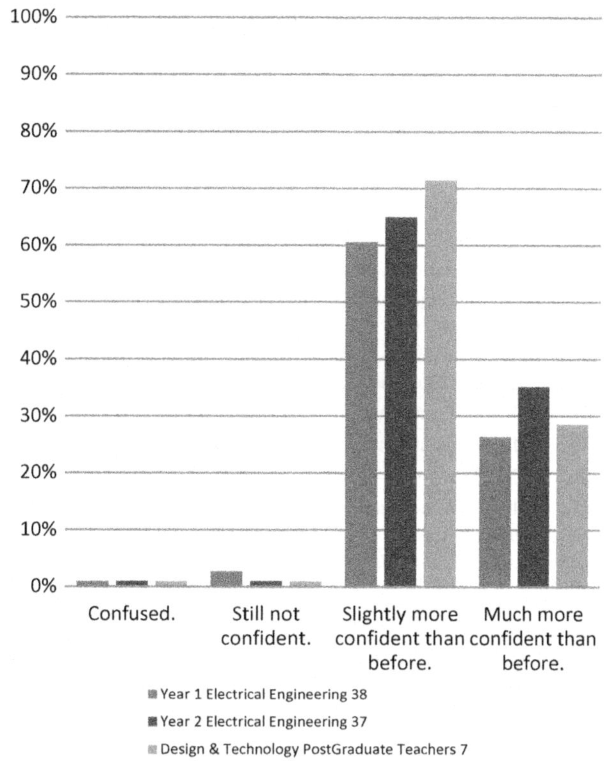

Fig. 6.10 Participants' rating of the utility of the video clip

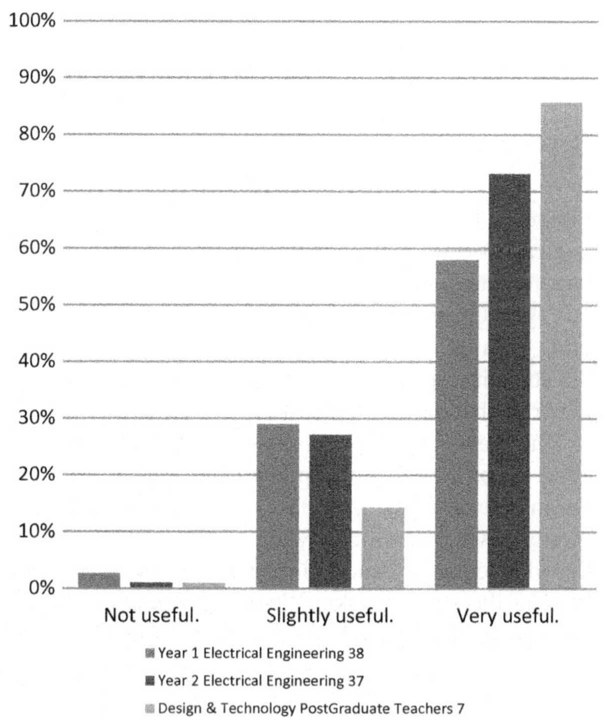

Table 6.4 Participants' ranking of resources after these were experienced first-hand

		Mean rank		
		Year 1 electrical engineering	Year 2 electrical engineering	Design and technology postgraduate teachers
Model description		N=34	N=37	N=7
A.	The hand-held cardboard model	1.75	2.58	3.00
B.	The animation, where the circuit components are superimposed on the graph (seen on video)	3.71	4.19	4.14
C.	The simulations from a software package, like Proteus	5.07	4.81	5.57
D.	The beer can model	4.21	4.09	4.00
E.	The comic strips	4.40	4.50	3.50
F.	The 3D wooden model	3.59	2.81	2.79
G.	The three-word sentences given as guiding steps	5.28	5.01	5.00
	P-value	0.000	0.000	0.056

detail. Participants were also in agreement that, such qualitative resources would complement other teaching methods, thus enabling students to learn more quickly and study more efficiently, while also helping them to be more motivated to pursue a better understanding of this the subject.

How this Might Be Used to Improve Teaching and Learning

Participant Learning Styles

Figure 6.11 shows a flowchart of how learning styles might be used to drive a pedagogical process. Subsequent to the measurement of learning styles, a teacher might plan to match and mismatch the forms of material taught and the assessment methods used. One might find that during the matching phases, the student is found to be more motivated and confident. While understanding the content with greater ease, the student might cover more grounds with respect to quantity and quality of the material absorbed and achieve better grades.

The mismatching phases might be used to actively drive the student beyond the zone of proximal development and into higher order developmental domains. During the mismatching phases, a student could be given the opportunity to become aware of which learning style can be considered more personally challenging. Such knowledge can be transformed into a positive experience by an effective teacher by helping the student acknowledge and devise strategies which can counteract any uneasiness experienced. This helps the student approach the behaviour of an expert

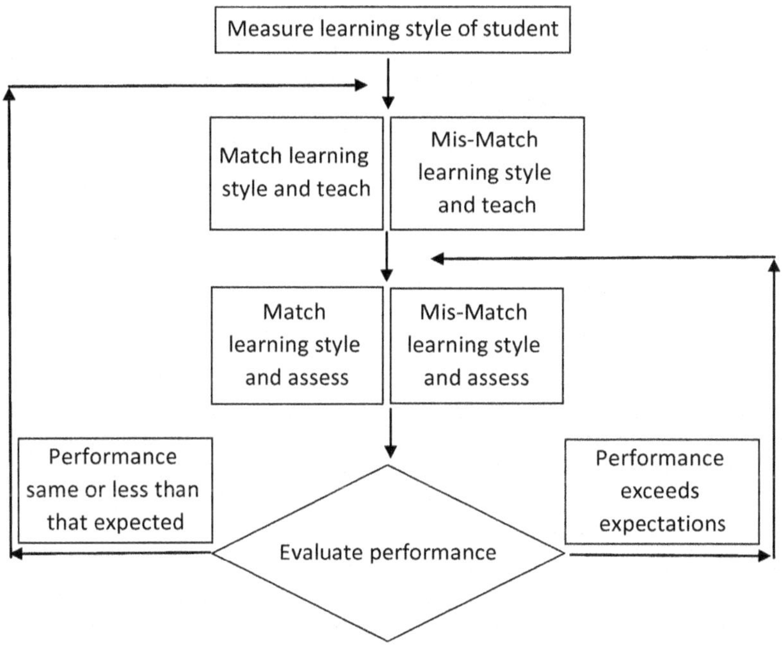

Fig. 6.11 Flowchart of learning styles

or a professional since it would help in recognising boundaries and plan how to overcome them.

This research indicates that actively targeting the audience's learning styles through the design of a teaching intervention does yield observable positive results in the learner's performance. The targeting of learning styles as a pedagogical design method can therefore be considered a valid approach towards improving learners' performance and better teaching approaches.

Current research arising from the domain of cognitive psychology informs that learning is dependent on the neural activity arising from diverse sensors within the body which fire signals towards the brain and not solely the mind. Outcomes from this research reinforce the dictum "*learning by doing*" or rather "*learning by experiencing*" and confirm that powerful learning occurs when a set of body senses work together, as in the case of active, sensing and visual learning styles, to contribute to the conceptual understanding of a phenomenon.

The methods used within this research suggest that assessment methods which are based solely on verbal propositions, rather than on visual imagery, could introduce a source of error when measuring students' genuine understanding of a concept. This may not simply be due to learning style alone, but also to constraint satisfaction. As Zhang (1997) claims, cognitive behaviour is much like constraint satisfaction, with many local minima, some of which may never be overcome without a change in representational forms. In electrical work, an exercise requiring students to write a verbal explanatory paragraph in order to describe a phenomenon may constrain their thinking less than an exercise in waveform sketching. With the former, students' thinking can be more abstract and therefore undefined and fluid than with the latter, as suggested by Lakoff and Johnson (1980) and Johnson-Laird (1983). A verbal paragraph may suggest that students have understood a concept when in fact they may have misunderstood the finer grain detail of the phenomena in question.

Discovery of Students' Conceptual Framework for a Phenomenon

Getting to know a students' conceptual framework for a phenomenon is important because it enables instruction to be based on the principles of constructivism. This entails the recognition of prior knowledge as a valid stepping stone towards the creation of new knowledge for the individual. This benefits the student by presenting new knowledge as familiar to the old, but it also benefits the teacher because it should take less effort to get the message across to the students.

Within this research, the discovery of a student's conceptual framework for a phenomenon was found to be better exposed when the method of inquiry involved interaction with a single individual for a stretch of time. Experience with groups has shown that it might prove difficult for a teacher to distinguish subtle elements in conceptual frameworks if students can only be interacted with as a group. Within this research the main source of data which fed the discovery of a conceptual framework was the written work submitted by the student and the conversations held on the topic in question. Careful scrutiny of formal and informal written work revealed interesting patterns with students' doodles proving to be especially helpful for tracing a probable line of thought. Body language such as hand gestures also played an important part in the detection of the elements, and links constituting the conceptual framework and verbal responses, including the audible mumbles, were key to build a coherent representation of the conceptual framework held by the student.

The formation of an accurate representation of a student's conceptual framework essentially entails the discovery of two features: (a) the elements constituting the content of the schemata and (b) the links in between such elements. A limited conceptual framework can be inferred not only by the direct evidence gathered on the topic but also by observing the behaviour of the student while problem-solving. Some symptoms indicating a limited conceptual framework included the following. Extracts from interview transcripts are given to manifest the symptoms noticed.

i Evasion of the problem. Within this research some participants admitted that when they did not know the answer to a problem, they tried to detect a pattern in the sequence of the test questions rather than pursuing the solution to the problem.

a) *if these ones of the graphs, had to be right after each other, I would have known the combination.*

ii. Attempt to recall information and doubting the recalled information when challenged.

a) *Well when I was at technical school, I used to remember the formulas of the time constant, etc. Now, there is much I have forgotten. you become very doubtful, … so much so that you start doubting even the simple things … sort of, you try to remember and see with what you can possibly grasp, … and you don't find much. I'm finding myself in the situation whereby I'm trying to look and grasp to the past, … and I'm not finding any lifeline… even the simple things are confusing me and making me doubtful.*

iii. Relying on dichotomous reasoning to suggest a solution and providing unelaborate explanations.

a) *Here, in this case I cannot recall the answer. So now, I will look at the graph. So in this case, I shall need the graph to reason things. I have two options, either charged or discharged, and I know that it is not charged, … so it must be discharged.*

iv. Displaying a resistance to consider all variables of the problem. Filtering and paying attention to those variables which are familiar thus evidencing a limited cognizance of problem landscape and consequently fitting their reasoning to the chosen variables only instead of taking a holistic approach to the problem.

a) *Discharged, charged. So, from the circuit I am saying that it charged. The graph is telling me that it discharged, but I know that it charged. I'm not even going to pay attention to the graph. I will focus on the circuit.*
b) *I reason it solely from the circuit, … it charged. I know that the graph is the opposite of charging, … so it's better if I don't use it, because normally when I focus on something, then I will change my reasoning.*

v. Not adopting a scientific approach towards finding a solution to the problem. Participants sometimes suggested approaches to problem-solving which defied the necessity of ensuring repeatability of outcomes or generalisation of explanations provided. They considered the circuits given as unrelated to each other and perceived the problems given as a multitude, rather than as variants of the same theme.

a) *This one is charged. No sorry, wait a minute, slowly, … this is discharged. The circuit is telling me discharged and the graph is telling me charged, but I'm going to tick discharged.*

vi. Display of limited analogical thinking powers and consequently functional fixedness. Disregarding the conditions or boundaries where an analogy becomes invalid or cumbersome.

a) *A capacitor is a tank – nothing else, it just stores.*

When a student was seen to display such symptoms, it was concluded that they were an indication of limits in his/her cognitive framework for the topic in question. Probing by questioning further enabled the researcher to track the roots of such behaviour and map the restricted number of schemata elements and links.

During the post-test participants' behaviour was seen to change. The observations presented hereunder were attributed to the development of a richer conceptual framework than that possessed before the teaching intervention:

i. Participants engaged in elaborate explanations involving all variables, not just the familiar ones. Familiar cases were also tackled in an elaborate manner even though usually the participant used recognition to reconfirm the reasoning provided.

   a. *So the capacitor here has discharged. Now point B. Here it was at zero volts. Then, ... I'll draw it. [see rough paper] Point B was zero volts and it went up to, ... 10 V. The graph is correct. I don't want to guess them or use the old method. That's why I am taking my time to reason out loud, even though I don't really feel I need to now.*

ii. The explanations given were consistent throughout the cases considered. While giving an oral explanation, participants were confident in their speech and their tone of voice was authoritative, sometimes even challenging back the researcher. They usually drew more than during the pretest, visualising and recording their thoughts on paper instead of simply verbalising them.

iii. Participants developed an eye for detail and exhibited metacognitive self-corrective thinking and action. Some participants evidenced profound understanding of the phenomena by pinpointing the more subtle behaviour of the circuits, sometimes even going beyond the information given by the graphical representations or analogies presented in the study. Such cognition was seen to promptly lead to further knowledge transfer and generalisation of concepts.

   a. *... so the capacitor is charging. Node B was zero volts at the start. The graph is incorrect. No wait a minute. [reasons things out] ... no, ... ok the graph is correct. It needs that vertical line from zero to 10, ... yes because now I'm looking for it, ... the story really starts from zero in the diagrams, ... so when I said zero, ... I automatically looked for the zero at the start of the graph, ... and because I did not see it, I jumped into the conclusion that it was incorrect.*

iv. Participants exhibited ownership of knowledge. Once they realised that they could juggle the information confidently, they switched from an impersonal mode of speech, '*the lecturer had told us that ....*', to a more personal one, '*now I understand that this is what happens ...*'.

Vigilance of such behaviour can give precious clues to a teacher about the state of the conceptual framework possessed by students. Being sensitive to such schemata usually leads to better teaching because the schemata inform the teacher about the students' prior state, and so the teacher can design learning pathways to 'fill in the blanks' in the schemata, reorganise the links within the schemata or augment and link more information into the schemata in a more deliberate way, thus increasing cognitive flexibility as defined by Spiro et al. (1988).

What Might Be Investigated Further

Recent perspectives of schemata or cognitive frameworks view them as processes that are dependent and grounded by properties of the human body and its external environment, that is, more dependent on perceptual aspects rather than just on abstract cognitive aspects. This gives rise to the notion of '*embodied cognition*', which makes the argument that just as the mind influences bodily actions, in turn, so do bodily actions influence cognition (Barsalou 1999, p. 613; Gureckis & Goldstone, op. cit.; Johnson 1987; Lakoff 1987; Lakoff and Johnson 1980). What type of experiences is possible, the significance of these to the individual and how the individual interprets them, understands them and reasons about them are integrally tied and dependent upon the character of the individual's bodily experience. This strengthens the very philosophy of design and technology because it places weight on the 'doing' actions as much as the 'thinking' actions, the argument being that the 'doing' feeds the 'thinking' just as much as the 'thinking' is necessary prior to the 'doing'.

Here is one example which the author has developed outside the focus of the doctoral research which capitalises on the finding that using the principle of embodied cognition improves learning.

When teaching about standing waves and the related terminology of nodes and antinodes in a physics class, the typical experimental setup for investigating such phenomena is usually small, and the nodes and antinodes are not directly visible but need to be inferred by placing tiny pieces of paper on a vibrating string and noticing if they fly off it or not. To make the experiment more perceptually based, the author used the setup shown in Photos 6.1, 6.2, 6.3, 6.4 and 6.5. This employed a home-made electronic module with a variable speed motor, a simple crankshaft

Photo 6.1 Variable speed motor with wire crankshaft

Photo 6.2 Fundamental or first harmonic

Photo 6.3 Second harmonic

Photo 6.4 Third harmonic

Photo 6.5 Fourth harmonic

made from metal wire and a rope having the thickness of a shoe lace. The experiment was conducted by having one student control the speed of the motor and another controlling the tension of the rope in such a way as to form different harmonics of the standing waves as shown in the photographs. The helper students developed 'a feel' for tuning the equipment to achieve a sustainable standing wave not just by the visible medium but also by the kinaesthetic pull of the tension in the rope and the speed of the motor. Moreover, onlookers usually asked the author's permission to 'catch a node' with their fingers, because they wanted to feel the 'relative lack of vibrations'.

Some examples where embodied cognition could be used to address other concepts include the following: (a) using a see-saw or simply holding weights in both arms to explain the concept of balancing a mathematical equation; (b) extension/retraction of the arms while performing a pirouette to explain moment of inertia; (c) alternating between a flat-foot walking posture and ballerina tiptoe posture to feel the effect of pressure and its dependence on area rather than force alone; (d) employing body partner physical fitness exercises to explain the action and reaction of forces, centre of gravity and stability; and (e) using marble machines or group activities to explain binary numbers and their manipulations such as shift, rotate and divide by n.

How Might Teachers Contribute to These Investigations

Teachers might find that adopting an embodied cognition approach for lesson planning yields positive learning outcomes for all students. Developing a pedagogy based on embodied cognition necessitates that the teacher considers how a concept can be made visible or tangible, that is, detectible by the human sensory system rather than just by mediating instruments. This can be achieved through designing a dynamic tangible model, adopting a real-life analogy or using the human body itself as an instrument. Students' conceptual understanding should be observed in a number of ways not just through the written verbal methods but also through sketches and bodily gestures. Video-capture techniques, or the employment of digital storytelling methods, could be particularly useful to record students' explanations of phenomena in order to discover an underlying conceptual framework since such methods allow for multiple modes of expression.
An electronic copy of the PhD thesis can be found at this URL: https://dspace.lboro.ac.uk/dspace-jspui/bitstream/2134/14637/4/Thesis-2014-Pule.pdf

Acknowledgement Acknowledgements for the support in my doctoral studies go to Dr John McCardle from Loughborough University, UK, Professor Liberato Camilleri, Professor David Zammit-Mangion and Professor Pulé from the University of Malta (UOM), The Department of Electronic Systems Engineering, the Faculty of Engineering and the Faculty of Education at UOM, together with ex-undergraduates from both these faculties, and Elspeth Musca.

References

Barsalou, L. W. (1999). Perceptual symbol systems. *Behavioural and Brain Sciences, 22*, 577–660.
Felder, R. M. (1988). Learning and teaching styles in engineering education. *Engineering Education, 78*(7), 674–681.
Goodhew, P. J. (2010). *Teaching engineering, all you need to know about engineering education but were afraid to ask*. Liverpool: The Higher Education Academy.
Johnson, M. (1987). *The body in the mind: The bodily basis of meaning, imagination, and reason*. Chicago: University of Chicago Press.
Johnson-Laird, P. N. (1983). *Mental models, towards a cognitive science of language, inference, and consciousness*. Cambridge, MA: Cambridge University Press.
Lakoff, G. (1987). *Women, fire, and dangerous things*. Chicago: University of Chicago Press.
Lakoff, G., & Johnson, M. (1980). *Metaphors we live by*. Chicago/London: The University of Chicago Press.

Lucas, B., Hanson, J., & Claxton, G. (2014). *Thinking like an engineer: How to cultivate engineering habits of mind in schools, colleges and universities*. London: Centre for Real-World Learning at the University of Winchester and Royal Academy of Engineering.

Meyer, J. H. F., & Land, R. (2003). Threshold concepts and troublesome knowledge 1: Linkages to ways of thinking and practising within the disciplines. In C. Rust (Ed.), *Improving student learning – Ten years on*. Oxford: Oxford Centre for Staff & Learning Development.

Pule', S. (2012). The formation of visual-verbal paired associates for resistor-capacitor series circuits. *Global Journal of Engineering Education, 14*(2), 143–154.

Spiro, R. J., Coulson, R. L., Feltovich, P. J., & Anderson, D. K. (1988). *Cognitive flexibility theory: Advanced knowledge acquisition in ill-structured domains*. Retrieved from.

Zhang, J. (1997). The nature of external representations in problem solving. *Cognitive Science, 21*(2), 179–217.

Chapter 7
Difficulties in Teaching and Learning Sectional Drawing in a University Based in the Eastern Cape, South Africa

Samuel Khoza

Abstract This chapter discusses the difficulties of teaching and learning sectional drawing in a university based in the Eastern Cape Province, South Africa. The study used both qualitative and quantitative approaches with subjects from the second- and the third-year lecturers and students engaged in Engineering Graphics and Design in the degree course.

The Questions I Asked and Why I Think They Are Important

Sectional drawing is one of the Engineering Graphics and Design (EGD) concepts taught in a university based in the Eastern Cape, South Africa. Students found it difficult to learn, and lecturers too had difficulties in facilitating the concept. In order to assist both the students and lecturers to successfully learn and teach sectional drawing, respectively, the following questions were asked:

1. *What difficulties do the student teachers in a university based in the Eastern Cape have in sectional drawing?*
   The above question was asked to both the students and lecturers engaged in EGD in order to determine how sectional drawing is learnt and facilitated. The answer to the question was supported by the application of both Vygotsky's theory of Zone of Proximal Development (ZPD) (1978), Piaget's theory of imagery and perception (1971), the Pedagogical Content Knowledge (PCK) theory that was coined by Shulman (1986), as well as the Purdue Spatial Visualization Test of Rotations (PSVT:R) developed by Guay (1976).
2. *What is the students' level of spatial ability application in sectional drawing?*
   According to Sorby (2003), students who learn EGD concepts need spatial visualization skills in order to successfully understand EGD. The above question was directed to the EGD students to diagnose their level of spatial skills. Both Piaget's

S. Khoza (✉)
University of the Witwatersrand, Johannesburg, South Africa
e-mail: Samuel.Khoza@wits.ac.za

PJ. Williams, D. Barlex (eds.), *Contemporary Research in Technology Education*, Contemporary Issues in Technology Education, DOI 10.1007/978-981-10-2819-9_7

theory of imagery and perception and the PSVT:R were used to assist in answering the above question. The PSVT:R contained three sections: development, isometric views, and rotations, and it is used to test for spatial skill. But the sections that were used in the study were only the isometric and the rotation views because they are relevant to the teaching and learning of sectional drawing. The two sections each contained 12 questions.

3. *To what extent do the students understand the uses and the application of line types in EGD?*
   According to Moolman and Brink (2010), there are ten different types of lines that are used in the EGD curriculum that also apply in schools. Out of these ten line types, seven of them are the main ones in EGD concepts with the other three mainly being the applications of some of the other types. Therefore the above question was posed to the students and lecturers engaged in EGD to know the extent to which line work is understood and used in sectional drawing. The classroom observation schedule was used to answer the above question. The observation schedule was adapted from the Community College of Aurora's Mentor Program Handbook and Staffordshire University's "Guidelines for the Observation of Teaching" (2000).
4. *How is the teaching and learning of sectional drawing occurring in the EGD classrooms?*
   In order to justify the methodology that were employed in the study, one of the ways to ascertain how sectional drawing is learnt and facilitated was to observe the teaching and learning practices of sectional drawing. The classroom observation schedule was used to get responses to the above question. The ZPD theory of Vygotsky was also used to interpret the learning activity in the EGD class during sectional drawing lessons.

The importance of sectional drawing is that it emanates from the development of visualization, which is a skill that is applicable in a lot of fields across the Science, Technology, Engineering, and Mathematics (STEM) areas (Nakin 2003). This is so because visualization is an essential skill in human intelligence, which helps an individual to easily learn geometry. On the other hand, learning and understanding geometry stresses visual thinking which is a skill needed in fields of aircraft maintenance and aviation. This is where individuals become able to manipulate objects and work with them. For the most gifted individuals, learning geometry requires symbolic knowledge, and the less gifted students require visual knowledge, which are all within the visualization skill. Acquiring such a skill supports entrepreneurship where individuals can use hand skills to make objects in technical, built engineering, processing, etc.

How I Tried to Answer the Questions

The study made use of both quantitative and qualitative approaches. The quantitative data was presented and analyzed statistically using the SPSS software. Likert scale questionnaires were given to both second- and third-year EGD students of a university based in the Eastern Cape.

Forty out of fifty EGD students studying toward a Bachelor of Education: Technical Education (BEDTEE) completed and returned questionnaires, which made it an 80 % response rate. There were 25 and 15 second- and third-year EGD students respectively. Questionnaires were used in the study to answer the question: *What difficulties do the students in a university based in the Eastern Cape have in sectional drawing?* The reliability of the questionnaire items was computed using Cronbach's alpha.

Qualitative research was used to obtain information through structured focus group interviews conducted with the 40 EGD students. Both the second- and third-year EGD students were given the PSVT:R to test their spatial visualization skill. The test was used to answer: *What is the students' level of spatial ability applications in sectional drawing?* Focus group interviews were used to answer: *What difficulties do the students in a university based in the Eastern Cape have in sectional drawing?*

Face-to-face interviews were also conducted with the two EGD lecturers. A non-participant observation schedule was used to observe the EGD lessons during the facilitation of sectional drawing. The purpose of the observation schedule was to get help in answering the questions: *To what extent do the students understand the uses and the application of various line types in EGD?* and *How is the teaching and learning of sectional drawing occurring in the EGD classrooms?*

Document analysis was also used in collecting data. According to Weber (1990), documents reveal what people do and what they value. This behavior of what people do occurred in an EGD classroom, so the data had strong validity. The researcher made use of previous assessment results, EGD models that are useful to sectional drawing, previous mark sheets, and course outlines of EGD second- and third-year students.

What I Found Out

Students first wrote the PSVT:R test before the commencement of the academic year to diagnose their spatial skills. Below are the results that show the students' performances.

Rotations sections are designed to "help students visualize the rotation of a three dimensional (3D) object" (Guay 1976). Table 7.1 shows that a lot of students performed poorly in the rotations section. The results showed that 72.5 % of students scored less than 60 % on the test which indicated that students' spatial skill was poor as they could not identify the correct rotated drawing which is one of the important EGD skills (Sorby 2003).

Analysis of an isometric section enables one to develop a skill to understand a three-dimensional (3D) view (Guay 1976). Table 7.2 shows that 62.5 % of students scored less than 60 % of the isometric section test that also indicated that students enrolled into the BEDTEE had poor spatial skill. The scores proved to be one of the indicators that students would have difficulty in learning EGD concepts because their spatial skill was poor (Sorby 2003).

Table 7.1 Students' performance on rotation sections

Score/12	% of test	Students N = 40	% of students
2	16, 67	6	15
3	25	4	10
4	33, 33	5	12,5
5	41, 67	5	12,5
6	50	4	10
7	58, 33	5	12,5
8	66, 67	3	7,5
9	75	3	7,5
10	83, 33	2	5
11	91, 67	2	5
12	100	1	2, 5
12 max		40	100

Table 7.2 Students' performance on the isometric section

Score	% of test	Students N = 40	% of students
4	33, 33	4	10
5	41, 66	5	12,5
6	50	7	17,5
7	58, 33	7	17, 5
8	66, 67	2	5
9	75	3	7,5
10	83, 33	4	10
11	91, 67	4	10
12	100	4	10
12 max		40	100

Questionnaire Results

Students doing EGD did not all have the same qualifications as shown below.

Table 7.3 shows that most students doing EGD had a Grade 12 qualification with only four having done EGD in Further Education and Training (FET). The FET N4 level is offered in technical colleges which are now referred to as FET colleges, and these levels are higher than Grade 12 where students study mainly engineering, mathematics, science, as well as EGD. NQF on the other hand is a South African qualification's framework that certifies a person if they have a relevant grade and/or results. This showed that students came into the program having done EGD at school and FET at college level.

Table 7.4 shows that 65 % of the students reported that they were familiar with EGD line types. According to Moolman and Brink (2010), knowledge of line types is key in EGD concepts; therefore, it was evident that the 35 % of the students who said they were unfamiliar with line types would probably find sectional drawing difficult.

Table 7.3 Students' highest school level

Students' highest school level	Frequency	Percent
Grade 12	36	90.0
FET NQF level 4	2	5.0
FET N4	2	5.0
Total	40	100.0

Table 7.4 Students' familiarity of technical drawing line work

Students' familiarity of line types *(N=40)*	Frequency	Percent
SA	11	27.5
A	15	37.5
D	12	30.0
SD	2	5.0
Total	40	100.0

Focus Group Interview Results

Focus group interviews were conducted with all the EGD students by the researcher. The focus group interviews were first transcribed; after which a sequence of recurrent themes was extracted. There were a total of eight groups of both the second- and third-year students doing EGD. The students' responses were presented as follows.

What Are the Difficulties in Learning Sectional Drawing?

The above question produced five themes. Two of the five themes are *lack of secondary schools' EGD background* and *sectional drawing being an abstract topic.* The majority of the students did do EGD at school level but lacked spatial skills that enabled them to learn sectional drawing as shown on the PSVT:R results. The reason why I only discussed two themes out of the five is because of the relevance they have to the study and the way they directly respond to the question under consideration.

Theme 1: Lack of Secondary Schools' EGD Background

A third-year student said "I did Graphics from Grade 10 but our school never had sectional drawing models of the exact types of sketches that we did."

The above response indicated that students did not have a full foundational background in sectional drawing. Even though most of them did EGD before, they came with a poor sectioning background. According to Brink et al. (2003), sectional drawing demands the basic knowledge and skills of EGD of Grade 9 curriculum, where graphic communication is done.

Theme 2: Sectional Drawing Being an Abstract Topic

A second-year student commented "Everything that I have drawn since the beginning of this topic is new to me, I have never seen them. I know that there is a piston but I have never seen one."

This was difficult for students who found sectional drawing irrelevant without drawing models. Therefore it was evident that most students' difficulties were also caused by the abstract nature of the topic.

The Extent to Which Students Understand the Uses and Application of Various Line Types

One of the themes that emerged from the above question was *knowing the importance of various line types*. The majority of the students knew the importance of various line types used in EGD with a few others not knowing their applications.

A second-year student said "I know that line-work is the basis of drawing but I really don't know why other lines are so important in sectioning."

The response above indicated that students knew the importance of line types, but they did not know why a variety of line types are important. According to Moolman and Brink (2010), line work enables students to communicate ideas graphically in engineering. This means that one cannot perform sectional drawing well if one does not apply line types correctly in order to describe hidden features in a sectional drawing.

Lecturer's Interview Results

The two lecturers' interviews gave responses to: *To what extent do the students understand the uses and the application of various line types in EGD?* and *How is the teaching and learning of the concept 'sectional drawing' occurring in the EGD classrooms?* Three themes emerged from the above question, one of which is sectional drawing is an abstract topic.

A second-year EGD lecturer said "I do not have drawing models to explain concretely in class, we rely on theory, which is not enough at all." The above lecturer's response clearly indicates that they found it difficult to teach sectional drawing and its concepts without sectional drawing models, which in turn made it hard for students to learn. Challenges of not having teaching resources and students not having learning resources did hamper their pedagogical activities.

Students' Classroom Observation Results

Not all of the third-year EGD students had drawing instruments. There were no teaching models of what the lecturer introduced, and the activities were very abstract. A lot of errors were spotted in students' tasks. However, the display of quality lines was of a high standard as compared to the second-year students. On the second-year class, the lecturer left the students unmonitored after presenting a lesson, and students saw that as a chance to do things other than drawing.

Lecturers' Classroom Observation Results

Illustrations of drawing principles were done through drawing diagrams. The findings imply that lecturers were unable to construct and provide alternative representations in the form of examples, illustrations, and analogies for students to understand better (Shulman 1986).

The lecturer in the third-year class used practical examples that enabled students to see the models they were drawing. On the other hand, students still found the practical examples used strange to them because there were components that they have never seen before. This showed that despite the lecturer having used practical examples, it was difficult for students to figure out how the object looked after being cut. This created problems for students when learning challenging sectional drawing concepts.

Document Analysis Results

Document analysis was used to help answer: *How is the teaching and learning of sectional drawing occurring in the EGD classroom?* as well as *What are the students' difficulties in learning sectional drawing?*

It was found that lecturers' subject files, which should normally contain formal documents like lesson plans and assessments, were also used as preparation files, which contained rough work, extra exercises, as well as any researched information related to the subject. In the second-year course outlines, sectional drawing was allocated three periods of facilitation per week, which lasted 90 min (30 min per period), but the lessons observed were only two in just 1 week, which were short of 30 min. No drawing models and study guides were available in the EGD laboratory. There were no display boards of previous drawings to motivate students. In the second-year lecturer's preparation file, sectional drawing assessment was done twice, and there was no evidence of assessment of sectional drawing in the third-year lecturer's file. The unavailability of course outlines for third-year students made it difficult for students to study further because they were not aware of the topics that followed. Also in third year, the assessment was frequent in that sectioning was spread across a number of topics like assembly, civil, and isometric drawing. The logic of topics from the second to third year was not easily determined.

The study has found that both the EGD lecturers and students have difficulties in teaching and learning sectional drawing. Lecturers found it difficult to teach without sectional drawing models, which made the topic too abstract for the students to learn and understand. Students knew some of the objects they drew by name, but have never seen these components in real life. The use of line types was also a problem during the observation period when learning sectional drawing through the production of sketches. Students were using wrong line types in situations where there were limitations of the uses of such line types. The assessment of sectional drawing was inadequate and not as frequent as indicated in lecturers' subject files.

How This Might Be Used to Improve Teaching and Learning

The findings of the study showed that the lecturers and the students had difficulties in teaching and learning sectional drawing without sectional drawing models. This made the sectional drawing an abstract topic to teach and learn. Incorrect application and misinterpretation of line types made it difficult for students to visualize a drawing. This may have made lecturers less willing to assess students adequately and often. I then saw it appropriate to come up with a teaching and learning model that I believed could assist in improving the teaching and learning of sectional drawing. This model presents a step-by-step approach that can be followed to assist both the teaching and the learning aspect of sectional drawing (Fig. 7.1).

The model presented in Fig. 7.1 shows that at the center of the learning experience, there should be an individual who understands all the EGD concepts, the curricular saliency, and the subject matter knowledge (SMK). The lecturer needs to be articulate in terms of linking components that are learnt through his teaching. But first as per the above model, the lecturer needs to diagnose students' (1) spatial skills in order to get the level of EGD knowledge that the students have. The diagnosis of spatial skills will ensure that proper preparation is made for the academic year because the lecturer will have an idea of students' prior knowledge. (2) Course outlines should then be given to students in order to allow them to study on their own before the introduction of topics and also to ensure that lecturers and students keep track of EGD concepts using these course outlines. This could then be followed by a thorough explanation of (3) line work, which forms the cornerstone of EGD. In the interviews that were conducted with students, they said they do know that line work is important, but they struggled to know which line types are appropriate. Moolman and Brink (2010) assert that the use and understanding of line

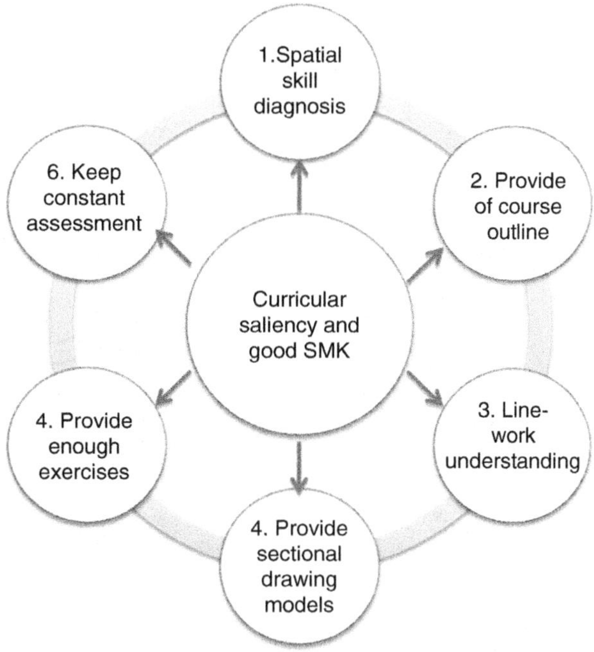

Fig. 7.1 Proposed model for the teaching and learning of sectional drawing

work are the co-determiners to a good sectional drawing. The fact that most students said that they knew line types but they did not know their applications proved to have been one of the students' difficulties in attempting sectional drawing. This difficulty was escalated by the abstract nature of sectioning due to lack of drawing models, as argued by Study (2001) that EGD and all its concepts need to be concretely learned and taught.

Therefore the above assertion revealed that for any EGD concept to be learnt successfully, line work is key since all drawings contain various line types that describe the features contained in each drawing. It is therefore important for the teaching and learning of line work to be emphasized during the initial instruction of EGD concepts before specific concepts are utilized. Teaching line types without more emphasis will likely not make any difference to alleviating the learning difficulty of sectional drawing.

When the line-work lessons are conducted appropriately and efficiently, (4) drawing models should then be provided in order to make learning concrete. This will also help ameliorate the confusion among students who said most drawings are based on new objects to them. After the collection of sectional drawing models, there should be more (5) exercises given to students. These exercises might not necessarily be those based on the models but other sectional drawing models that might be available to lay a foundation to develop visualization skills when an object is cut/sectioned. Thereafter, (6) assessment of these exercises should also be carried out to provide feedback to the students so they will know how they progress and maybe need to redo some of the tasks. All of these steps need a lecturer who knows

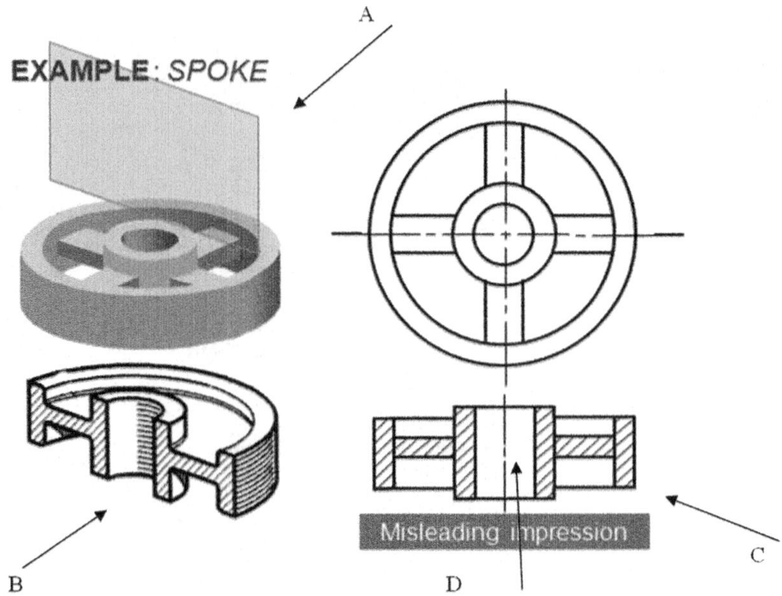

Fig. 7.2 A spoke (Moolman and Brink 2010)

his subject matter and who has good pedagogical knowledge to link concepts and make his students know and understand the reasons behind studying sectional drawing and how sectional drawing links with other concepts beyond the university level.

Figure 7.2 is a prime example of sectional drawing that EGD students had to answer in assessment activities. The errors that students made were more of the "principle mistakes" which clearly showed that their completion of sectional drawing was technically wrong in the application of line work. The fact that most students had not seen a real spoke and many other objects that they came across in EGD classes affected the manner in which students approached the examples.

In a typical sectional drawing question, drawing A is the principal view that a question emanates from, and it is represented in a 3D format and a question would be phrased as follows: *Draw a full sectional front view of a spoke*. This can easily be translated as: *Draw a front view of a spoke when it is cut*. Therefore what students need to do is to imagine the inner features of the drawing. The imagining process is basically a synonym for visualization, which in addition needs a good understanding of line work for the drawing to be successful. The success of students in visualizing the drawing would assist in the production of a correct solution. And also if the model for the spoke and other unfamiliar models were available, it would have been much easier for students to understand and see the result of the cutting action in order for the solution shown in B to be drawn.

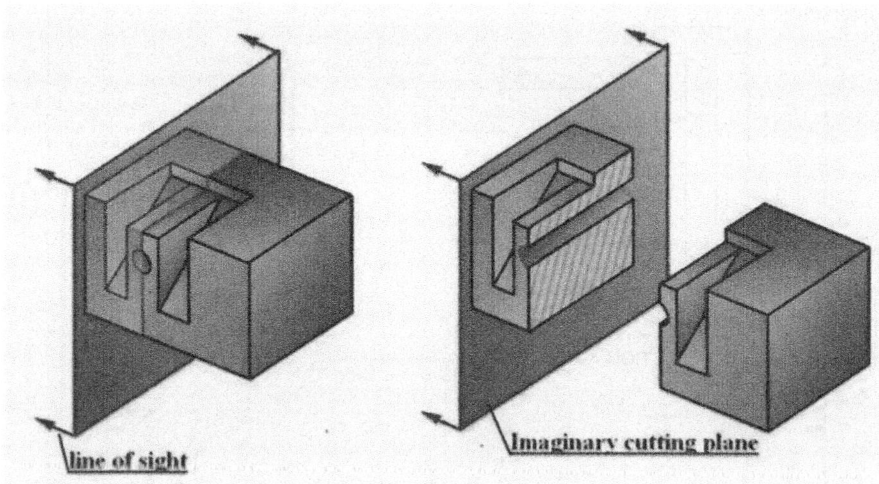

Fig. 7.3 An introductory sectional drawing model

Teaching Methods in Sectional Drawing Lessons

Teaching methods of sectional drawing are affected by many contextual factors that lecturers mentioned in the findings of this study. Lecturers are most likely to be unable to employ various assessment strategies because their knowledge of the subject matter is affected by the fact that they do not have relevant material, drawing models, and workbooks to facilitate learning. Since the unavailability of resources was what lecturers complained about, then their teaching of sectional drawing would not be effective. Therefore the suggestions for the best teaching methods of sectional drawing are drawn from the model in Fig. 7.1.

Single-Component Sectional Drawing

The example only requires a sectional solution of one view, and the cutting plane is easily identifiable. A drawing model and a good understanding of line work are required to visualize the cutting action. The "line of sight" indicated identifies the cutting plane, and it further shows the two parts that are a result of cutting. This kind of example ought to be typical of the ones used during the primary stages of sectional drawing introduction, followed by the one shown in Figs. 7.3 and 7.4, which does not show the cutting plane using the "line of sight."

Figure 7.4 will require deep knowledge about line work in order for students to figure out how the drawing looks inside so they can provide a solution. It is a bit more difficult than the Fig. 7.3 model because it does not show the "line of sight" (Fig. 7.5).

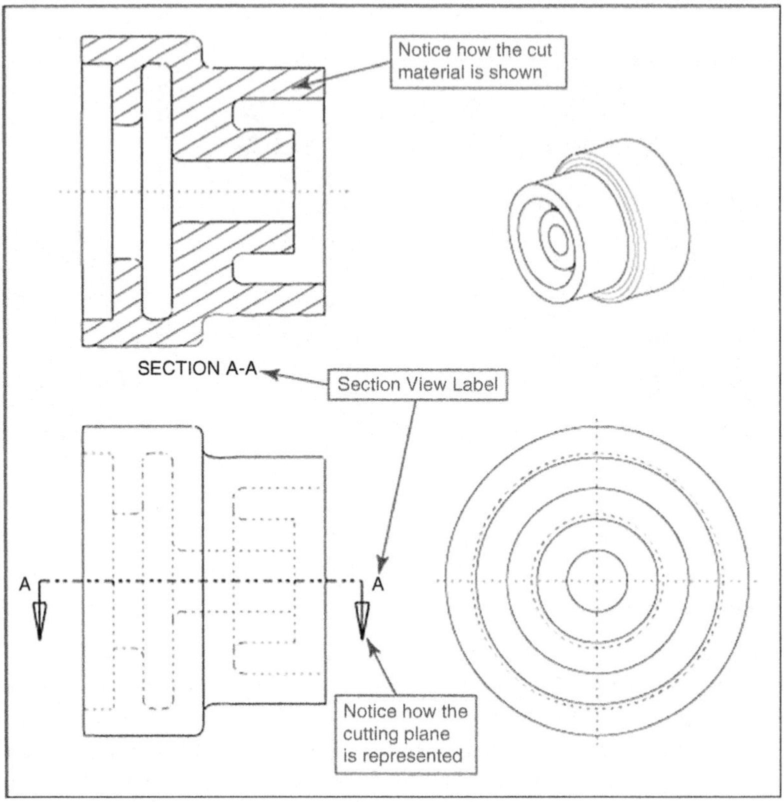

Fig. 7.4 A secondary sectional drawing model

Figure 7.5 is called an assembly drawing, and it is much more complicated than the ones shown in Figs. 7.3 and 7.4. It is a multi-component type of a model. Students will provide a solution to this kind of drawing only after they understand the working principles of Figs. 7.3 and 7.4. It requires good spatial skill. It requires students to visually assemble parts and later section them to accurately depict each one of them. The main communication and interpretation skill in interpreting the above drawing is line work.

Other Areas Related to Sectional Drawing Research

Having looked at the above examples, it is evident that spatial skills play a key role in sectional drawing and other EGD concepts. The recommendation therefore is that students' spatial skills need to be diagnosed at an earlier stage in order to allow the planning and facilitation of sectional drawing. It is also recommended that the admission criteria in South African universities emphasize the secondary schools'

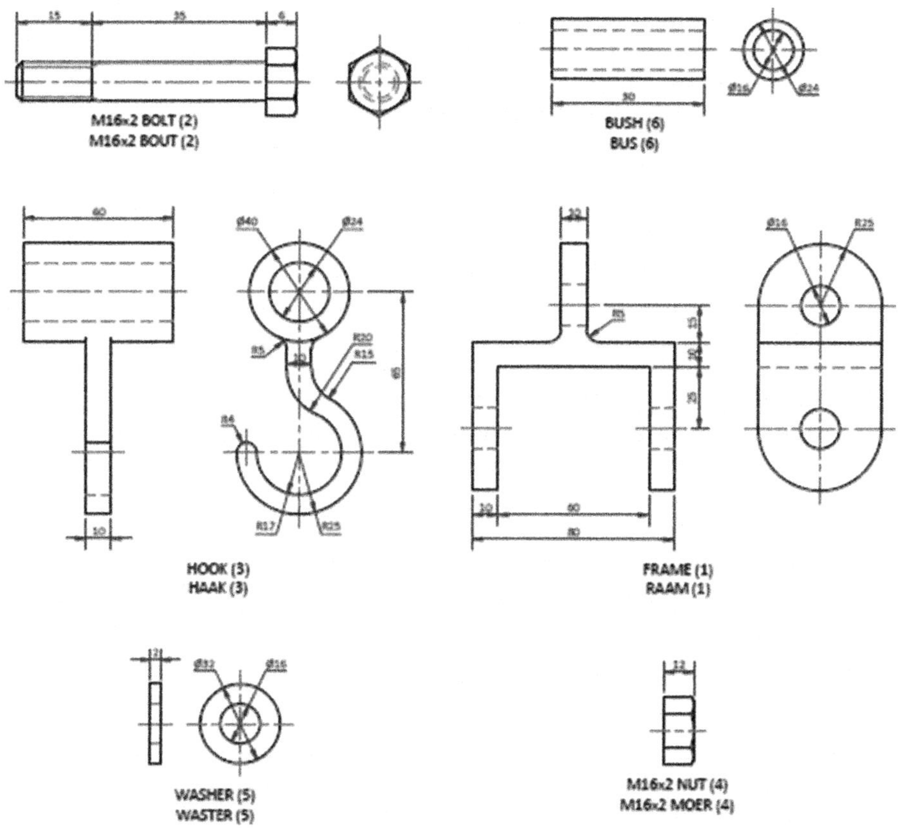

Fig. 7.5 Example of the assembly drawing

background in EGD. Alternatively spatial skills' diagnosis ought to be a benchmarking test for anyone studying towards EGD and, drawing models should be provided for the concrete studying of sectional drawing. Subjects like mechanical, civil, and electrical technologies, which are mainly paired with EGD, should allow students to make drawing models that they will use in the EGD classes.

Piaget (1969: 24) says spatial skills are developed in three stages: (1) topological skills, (2) visualizing 3D objects, and (3) visualizing the concept area. These skills are acquired according to the way a child matures. Topological skills are primarily two dimensional, and with these skills, children are able to recognize an object's closeness to others. The second stage involves visualizing 3D objects and perceiving what they will look like from different viewpoints (Piaget 1969: 26). In the third stage, people are able to visualize the concepts of area, volume, and distance in combination with those of translation, rotation, and reflection (Piaget 1969: 29). The above skills are practical in nature and need concrete learning and teaching to develop. Therefore the unavailability of models makes the teaching and learning of EGD and all its concepts difficult.

On the other hand, Vygotsky argues that development is not a cognitive process that occurs inside a person's head and is separate from the external world in which people live (Vygotsky 1978). According to Vygotsky, the development of a child is not only cognitive and biological, as Piaget elaborated, but the development of the child is also social, that is, it is also influenced by what surrounds him. To Vygotsky, teaching and learning are an integrated process in human development. During the teaching and learning process, students are expected to actively participate in their own learning through the use of language and interactions with their colleagues and instructors. Therefore for the teachers in secondary levels, they need to inculcate the practice of collaborative learning among learners. Social interaction can therefore be encouraged by allowing students to learn among themselves in groups and come up with models that they can understand, preferably the models that they will have made in their workshops which they can use for sectioning.

However, subjects like EGD should be taught more practically, using drawing models in order to create a clear picture of what is being produced during the drawing process. This practice should also be carried over to universities so that technical subjects, even subjects like mathematics and sciences, start to be approached differently.

What Might Be Investigated Further

EGD is one of the subjects in a technical field with narrow job opportunities. Concepts like sectional drawing are also difficult to apply everyday, but the skills that are developed in mastering sectional drawing are applicable in a variety of fields. Since this study has found that spatial skills are crucial in mastering sectional drawing and other EGD concepts, the skills that enable students to master sectional drawing are applicable in other fields like mathematics, sciences, and built environment. The concept of line work that was revealed as a basic need to learn EGD helps in understanding geometry, line graphs in physical science, as well as drawing house plans in built environment. Therefore in order to curb the continuing challenges that exist in learning sectional drawing, it is important for students' spatial skills be diagnosed at an earlier stage before the commencement of the academic year in order for such spatial skill challenges to be addressed on time. This will assist the lecturers and curriculum planners to plan ahead and put relevant measures in a form of a bridging course that will bring the students' level of drawing understanding to a required level.

How Teachers Might Contribute to These Investigations

To curb the challenges that students face in spatial skills in tertiary-level engineering courses, teachers at school level need to take the initiative to develop spatial skills in learners at lower levels of secondary education through sketching. Sketching is one of the basic exercises that learners can be trained in producing simple drawings in subjects like life sciences (in anatomy section), mathematics (in geometry and planes), physical science (in line graphs), technology (in design processes), social sciences (in geography), etc. This will help learners to have a better understanding toward the STEM subjects, which over the years have been a challenge in South Africa's education system. Therefore, teachers at school level should also diagnose and focus on spatial skills at an earlier stage so that they can plan their work and select relevant drawing models that will ease learning and teaching.

A copy of this thesis is available at http://tkplib01.tut.ac.za/record=b1306933~s13.

References

Brink, C. G., Gibbons, P. J., & Theron, A. E. (2003). *Engineering drawing N1*. Johannesburg: Heinemann.

Guay, R. B. (1976). *Purdue spatial visualization test*. West Lafayette: Purdue.

Moolman, C. L., & Brink, C. G. (2010). *Engineering drawing N3* (3rd ed.). Johannesburg: Heinemann.

Nakin, N. (2003). *Creativity and divergent thinking in geometry education*. Unpublished Masters Dissertation. Pretoria: University of South Africa

Piaget, J. (1969). *The language and thought of the child*. Cleveland: The World Publishing Company, Meridian Books.

Piaget, J. (1971). The theory of stages of cognitive development. In D. Ross-Green et al. (Eds.), *Measurement and Piaget: Proceedings of the CTB/McGraw-hill conference on ordinal scales of cognitive development*. New York: McGraw-Hill.

Santos, J. R. A. (1999). Cronbach's alpha: A tool for assessing the reliability of scales. *Journal of Extension, 37*(2), 1–5. Available from http://www.joe.org/joe/1999april/tt3.html

Shulman, L. S. (1986). Those who understand: Knowledge growth in teaching. *Educational Researcher, 15*(2), 4–14.

Sorby, S. A. (2003). *Introduction to 3D spatial visualization: An active approach*. Clifton Park: Thomson/Delmar Learning.

Study, N. E. (2001). *The effectiveness of using the successive perception test I to measure visual-haptic ability in engineering students*. Doctoral dissertation, Purdue University. Retrieved from Proquest.com

Vygotsky, L. S. (1978). In M. Cole, John Steiner, Scribner, & Sauberman (Eds.), *Mind in society: The development of higher psychological processes. LS Vygotsky 1896–1934*. Cambridge: Harvard University Press.

Weber, R. P. (1990). *Basic content analysis*. Newbury Park: Sage Publication.

Chapter 8
Let's Get Kids Talking in Technology: Implications for Teachers

Wendy Fox-Turnbull

Abstract Classroom conversations are core to establishing successful learning for students. This research explored students' conversations in technology education in the primary classroom and suggests some of the implications for teaching and learning. It used qualitative methodology which paid particular attention to the social nature of the classroom. Participants took their own photographs which were used in conjunction with a range of interviews with participants and teachers. Students' work samples were also used to develop a rich description of classroom conversation in technology.

The Questions I Asked and Why I Think They Are Important

Classroom conversations are core to establishing successful learning for students for two main reasons. The first, dialogue between teachers and students, assists teachers by giving students insight into their thinking and understanding. This enables teachers to adjust planning and teaching to meet specific needs of their students. The second is that through engagement in dialogue with peers and teachers, students are able to expand their understanding and knowledge. This research explored the use of talk in technology education in primary classrooms and the implications for teaching and learning.

The aim of this research was to understand and describe the role talk plays in learning technology. The analysis of conversation transcripts, students' autophotographs and observations of behaviour were used to describe and analyse the nature of classroom talk.

In this chapter classroom talk is considered from two perspectives: strategy and knowledge. Strategy refers to the strategies used to ignite and facilitate the conversation in the study. Knowledge refers to funds of knowledge, learning area knowledge and technological content knowledge which contribute to students' technology literacy.

W. Fox-Turnbull (✉)
University of Canterbury, Christchurch, New Zealand
e-mail: wendy.fox-turnbull@canterbury.ac.nz

PJ. Williams, D. Barlex (eds.), *Contemporary Research in Technology Education*, Contemporary Issues in Technology Education, DOI 10.1007/978-981-10-2819-9_8

Main Question

What is the nature of conversation in technology education?

Subquestions

1. What types of conversations enable students to participate in collaborative technological practice?
2. How do students' prior and concurrent experiences influence their technological practice?
3. What happens in the classroom to increase the likelihood of students deploying knowledge and skills from other areas into technology?
4. What insights into technology education can be gained through an analysis of students' conversations with their teachers and peers while participating in technology education?

This study advances research in the area of learning in technology by studying students from two primary year levels working in the same or very similar technological practice. This allowed insight into how previous experiences, background and culture impacted on and contributed to students' understanding in technological literacy and practice and the types of talk that facilitated this process.

The findings of the study are useful and exciting because they help us understand how students learn in technology education. The study develops current understanding of the nature of talk and the role it plays in learning technology. It also presents new findings on the impacts that cultural knowledge and skills from home and community bring to technology. It also challenges existing findings on students' ability to transfer knowledge from one curriculum area to another.

How I Tried to Answer the Questions

This was a qualitative study which paid particular attention to the social nature of the classroom. In the study I interpreted the data to identify detailed aspects of the nature of classroom talk in technology. To do this I spent many hours in the two classrooms, one Year 2 class with 6- and 7-year-olds and one Year 6 class with 10- and 11-year-olds, over the period of a year, during the delivery of two technology units, each involving the planning and implementation of a different predetermined whole school theme. I took observations and oral recordings, interviewed students and teachers and gathered teachers' planning and students' work samples to develop a deeper understanding of the nature of classroom talk in technology education. The study took place in an urban New Zealand primary school.

In this research, the culture of the classrooms and the particular groups of students being studied were clear foci points. My role was clearly understood by all participants and I was present in the classroom during data gathering. As a registered teacher myself, I was able to quickly build a rapport with the students and establish myself as a teacher. The students' ability and willingness to tell their stories and share their ideas of technological practice with their peers, their teachers and me depended on, among other things, the culture of their classrooms. Technology education is a holistic and contextualised curriculum and therefore fitted well with the research methods I selected. I interviewed students initially and then became fully immersed within the culture of their classrooms. During the first unit 'futuristic travel' (Round 1), I assisted the teachers in the unit implementation. The second unit, 'props for the school production'(Round 2), was taught later in the same year and was when most of the data gathering occurred, as the students knew me from our previous work together.

In this study, I was a participant-observer. This meant I took a role in classroom proceedings while observing. Wolcott (1988) suggests this is an important way to gain information in this type of study. Taylor and Bogdan (1998) suggest that participant observation is particularly suitable within the natural classroom setting. I stayed with the participants for a substantial amount of time to reduce the effect I as a researcher had on the participants.

Stimulated recall using autophotographs was one of the research tools employed in this research. The participants were taught how to take photographs on digital cameras in Round 1 and given disposable cameras in Round 2 to record their own technological practice. Photographs were used because they allowed students to capture a specific moment or activity. The term autophotography has been used throughout this study to describe the process of self-generated photographs by participants. The photographs generated by the students were then used to stimulate discussion about technological practice. Disposable cameras were used because it enabled the researcher to give every student in the class their own camera. They were relatively inexpensive, sturdy and easy to use. Also students were not able to delete photographs taken.

During the analysis phase of the study, open coding was used. All interviews were audiotaped and then transcribed, and the participants recorded photographic evidence of their technological practice was added. Detailed anecdotal observation notes were taken as students worked.

Systematic and meticulous organisation of the data was required. The steps used for data analysis in this study follow the process suggested by Lichtman (2006) and included the following:

Step 1 Initial coding and recognition of some central ideas from the raw data
Step 2 Revisiting initial coding, refining and modifying where necessary
Step 3 Developing an initial set of categories or central ideas
Step 4 Modifying of initial list after some additional rereading
Step 5 Revisiting categories and subcategories
Step 6 Moving from categories into concepts (themes and perspectives)

Broad conversation categories were identified based on the source and purpose of the conversation, how and why the conversation occurred. Initial analysis of stimulated recall conversations led to the identification of four significant unit stages: character and function, planning, mock-up and further analysis lead to the identification of the four key elements of conversation, funds of knowledge, making connections, management of learning and technological knowledge. Identification of the stages and elements led to the development of the conversation framework which facilitated in-depth analysis of conversation.

What I Found Out

The research findings show that classroom conversation in technology is situated within three themes and occurs from two perspectives. Figure 8.1 shows three conversation themes related to the perspectives of learning they offer, conversations with a strategy perspective on the left and those with a knowledge perspective on the right. Each of the two identified perspectives of conversation also has a number of aspects, also identified in Fig. 8.1. The three overarching conversation themes or purposes of conversation undertaken by students and their teachers are deployment, conduit and technology knowledge as seen in Fig. 8.1. The three themes occurred through two different perspectives: strategy and knowledge. The three conversation themes worked together rather like a set of cogs with the conversations from strategy perspective, acting as a 'conduit' between the two *knowledge* themes: deployment and technology knowledge. This is illustrated in Fig. 8.2 which demonstrates the interconnected nature of the themes.

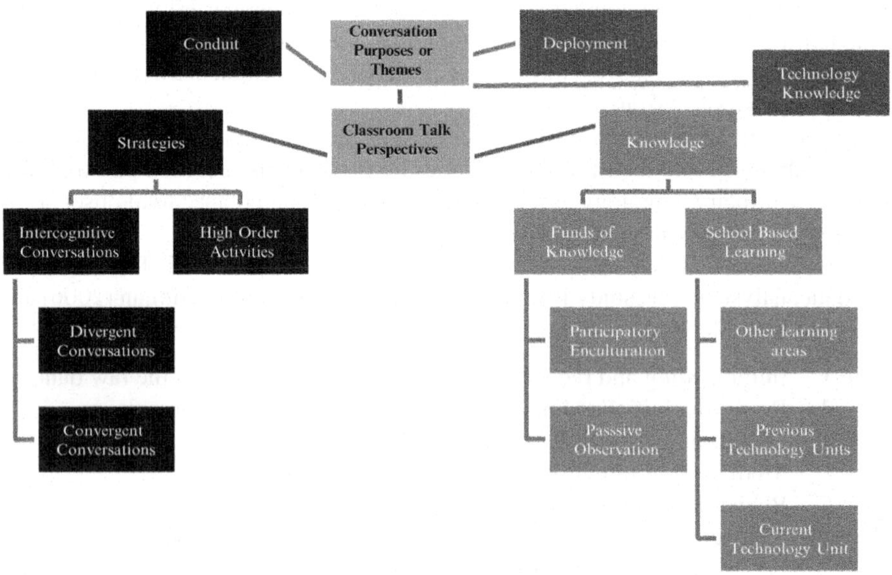

Fig. 8.1 An overview of key finding about classroom talk

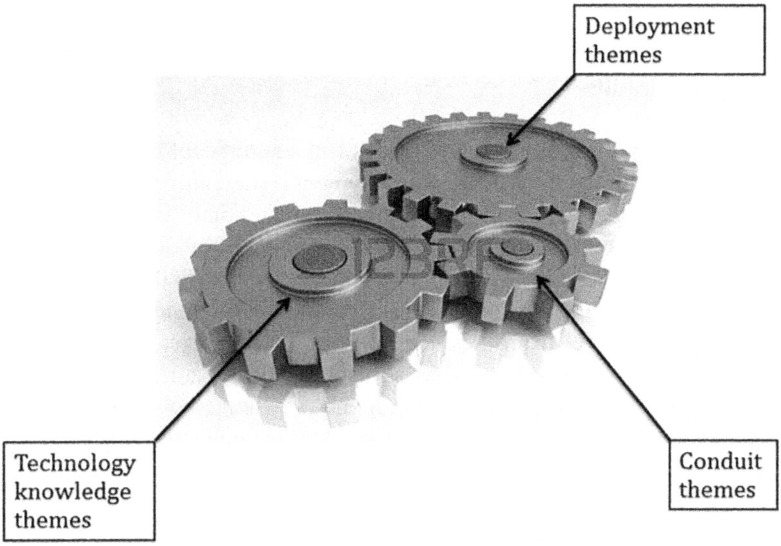

Fig. 8.2 The interconnected nature of emerging themes of conversation

Without the conduit themes, the deployment and technology knowledge themes remain unconnected and do not inform one another.

Themes

The first theme 'deployment' described knowledge and skills brought by students to their technological practice and therefore occurred through a knowledge perspective. Deployment themes identified knowledge and skills students deployed to facilitate their understanding of and learning in technology. They were sourced mainly from students' funds of knowledge and the links and connections they made before and during their current school-based learning.

The third theme, 'technology knowledge', was the second from the knowledge perspective and showed the exact nature of technology learning obtained by the students though the bringing together of the first two themes. The technology knowledge conversations were conversations where students demonstrated technological knowledge and skills in relation to their current project. Knowledge themes emerged from a synthesis of the deployment and conduit themes, discussed below, and evidenced students' understanding and learning of technological knowledge and skills and how they made connections to prior learning in technology.

The second theme 'conduit' described techniques and strategies used by teachers and students to maximise learning opportunities and acted as a conduit between other knowledges and technological knowledges. These conversations come from a strategy perspective. Conduit themes were identified from conversations facilitated and undertaken in the classroom based on management of learning, resources, time and behaviour.

Perspectives

Knowledge and Skills

The study found that students' conversations had a significant impact on their practice in technology. In the first theme – deployment – students come to their technology projects with significant knowledge from their home and cultural funds of knowledge by making connections to prior and current school-based knowledge. They therefore deployed knowledge and skills to contribute to their learning in technology, some knowledge they already possessed and brought to their practice without specific prompting from teachers. At times teachers explicitly drew on knowledge they knew the students had. This knowledge came in a range of forms and types and included not only direct content knowledge but also process knowledge and knowledge about ways to behave, for example, strategies for working collaboratively.

Funds of knowledge, knowledge drawn from home and community, were learned through two different methods. The first was passive observation in which learning occurred by watching without interacting such as watching TV or movies or reading a book. This was exemplified by Minnie who was able to use knowledge from a song she knew it assists her recognition of a picture of a waggon. 'Oh, it's from the olden days, a cart or something. Probably [used] like a hundred years ago or sooner, like. There's that song, Little House on the Prairie'.

The second method of obtaining funds of knowledge, participatory enculturation, occurred when students were actively involved in gaining new skills and knowledge. This was exemplified by Ellis and Anne who compared the process of stuffing the mock-up fish they were constructing in order to get a three-dimensional effect with the process of gutting and/or filleting a fish, which both children had experienced at home. Ellis had been salmon fishing with his grandfather and Anne with her immediate family. Ellis suggested that rather than removing salmon flesh from the fish, they were, in fact, adding to the fish. Anne agreed but used the more general term 'meat' rather than salmon.

Ellis Yeah, like we're actually putting all the salmon into the fish.
Anne All the meat into the fish and not all meat out of the fish.

Evidence of students' learning in technological knowledge was also evidenced in the third theme: knowledge. As the students worked through their technological practice, they evidenced learning of generic technology knowledge and skills, such as understanding the characteristics of technology, developing a brief and drawing and constructing technological outcomes. The nature of talk during students' technological practice altered as the students worked through different stages of their practice. In the early stages, the students were engaged in finding out about props in general and then more specifically 'their' props. Subsequently, their conversations changed to incorporate design and construction skills. Throughout, students were involved in talk with their peers and also with their teachers, at times collaboratively and at times one to one.

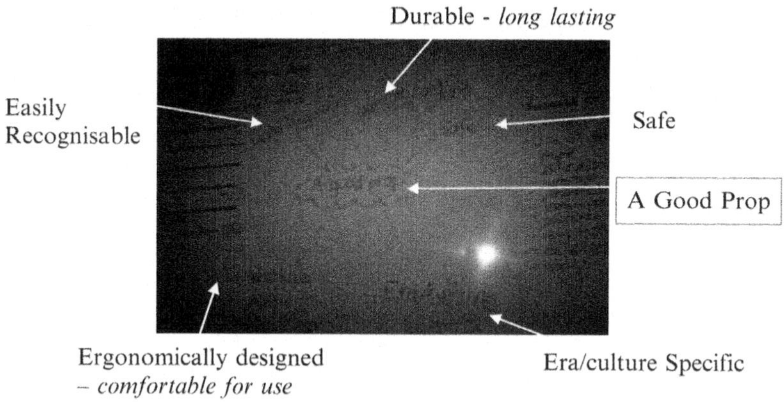

Fig. 8.3 Mandy's authophotograph of the initial class brainstorm with author added annotations for clarity

Strategies

All talks identified within the second theme 'conduit' were conversations used to facilitate and transfer learning to build understanding to technological practice. They therefore took a strategy perspective. Conduit conversations assisted students in recognising the relevance of prior knowledge and learning and gave students ample opportunities to explore, talk about and use pre-existing knowledge to enable this process. Talk in the conduit theme also included the implementation of teaching and learning strategies used to assist students' learning, managing their behaviour and resources. By explicitly drawing students' attention to potential sources of knowledge, teachers assisted deployment of this knowledge to learning in technology. Learning was facilitated through the careful implementation of planned and focused activities which enabled students' engagement in the synthesis, analysis and evaluation (Bloom 1956) of a new material. This study demonstrated that students' technological knowledge, skills and outcomes were enhanced through these planned learning activities and strategies. Two such strategies occurred in the Year 6 class, both illustrated below by Mandy's autophotographs. The first (Fig. 8.3) demonstrated brainstorming as a strategy to assist students' understanding of their topic. Mandy was able to tell me this about the photograph 'that was when we were thinking about a good prop, and we had to make it durable, safe, easily recognisable, ergonomically designed and specific to the era or culture'.

The second strategy illustrated in Fig. 8.4 is a PCQ chart, in which the students needed to identify the pros, cons and questions about, first, existing props and, second, their intended designs.

PCQ Idea:________________________ **Names:**________________

Pros	Cons	Questions
List all the benefits, strengths, plusses, advantages of an idea from as many points of view as possible.	*List all the negative aspects, contra ideas, disadvantages, weaknesses of an idea from as many points of view as possible.*	*Offers an opportunity to questions, curiosity, probing and 'what ifs'. 'I wonder…' 'What if…' or 'It would be interesting to know…'*

\canterbury.ac.nz\systems\Staff2Homes\wkt26\My Documents\PhD\PHD Data\Field Work\PCQ Template.doc

Fig. 8.4 Teacher template of the PCQ chart

Strategy Perspectives

Elements of conversation identified in the study indicated that students' conversations had a significant impact on their practice in technology. Of the three major themes of students' conversations, the conduit theme included strategies for learning to talk to advance thinking and strategies implemented by the teacher to assist students' higher-level thinking. Both impact on teaching and learning in technology.

From a strategy perspective, the conduit conversation theme was used to facilitate and transfer learning and understanding to students' technological practice. Conduit conversations included conversations and teaching strategies that assisted students' recognition of the relevance of prior learning. An implication for teachers is that students need to be given ample opportunities to explore, talk about and use pre-existing knowledge. Conversations in the conduit theme do this by acting as a pathway between the first theme, knowledge with the potential for deployment within technological practice, and the third theme, technological knowledge and skills. Conversations in the conduit theme also include the implementation of teaching and learning strategies used to assist students' learning. They assisted students by teaching them how to engage with their peers and by explicitly drawing students' attention to potential sources of knowledge. Learning can be facilitated through the careful implementation of planned and focused activities which enable students' engagement in the synthesis, analysis and evaluation (Bloom 1956) of new materials. This study demonstrated that students' technological knowledge, skills and outcomes were enhanced through these planned learning conversations and activities and strategies.

Teaching Strategies

The following sections illustrate a number of successful teaching strategies that can be used in technology. These are only a range of suggestions and the reader may have equally valid alternative suggestions. Using the context of this study, imagine teachers Fleur, working with 6-year-old, Year 2 students, and Clara, a teacher working with Year 6 (10-year-olds), asking their students to design and make props for their upcoming school production.

Through dialogue with each other, students were able to take knowledge and skill development further than they would have been able to do individually. This was exemplified by Rex (aged 6) who early in the study identified that working in his group was difficult but in the final focus group interview stated that by working together the group he had achieved more than he could have by himself. This has important implications for planning and teaching in technology. Talk is a vital component of learning. Teachers need to plan for and teach students to talk constructively, using debate and discussion as a tool for advancing thinking and understanding. During implementation students also need to be taught how to listen to and accept others' ideas without necessarily agreeing with them. Teachers also need to assist students to understand that, although their own ideas are not always accepted, their contribution may be still important because conflicting ideas and opinions force all members of the group to question and justify their decision making, thus making stronger connections to key concepts and knowledge.

Intercognitive Conversations

Intercognitive conversations describe a situation within which all participants learn through the talk and associated reflections. When participants are learning in, and about, a common context and engaged in constructive talk or dialogue, they actually assisted each other and advanced their own knowledge in and about technology. Debate, argument and/or disagreement also assists students' understandings in technology, but only if and when participants are open to change and new ideas. In situations where conflict arises, and because in technology students are often developing one outcome per group, they have to find a single solution, which means either acceptance of others' ideas or reaching a compromise.

In order to facilitate intercognitive conversations within their classroom, Fleur and Clara set up a classroom culture in which the students did not raise their hands to answer teacher questions. The students were taught to think independently, discuss, question and challenge their own and others' thinking without attacking or experiencing feelings of being attacked. They were also shown how to let go of

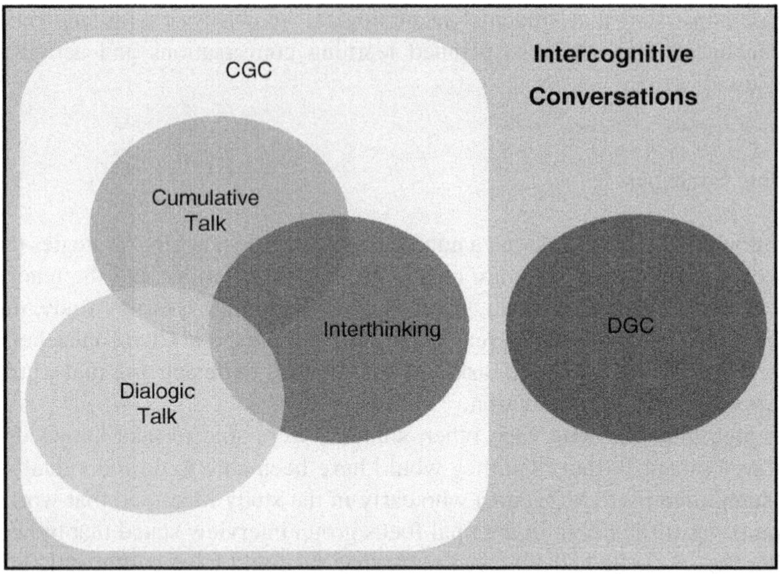

Fig. 8.5 The relationship between convergent and divergent growth conversations within intercognitive conversations

some ideas, to be open to the opinions of others and to alter opinions as new information come to light. When this occurred convergent growth conversations (CGC) happened with all participants moving to new understandings within the same context.

Intercognitive conversations are not restricted to students-students conversations. In this study while listening to the talking partner discussion, both teachers gained insight into how their students thought about technology and gained new knowledge about students' learning in technology. This type of conversation is characterised by divergent cognitive growth for all participants and is called divergent growth conversations (DGC). The relationship between the two is illustrated in Fig. 8.5.

Teacher Knowledge and Understanding of PCK and Students' Higher-Order Strategies

Strategies used in the study to enhance students' deeper understandings of technological process and the context of learning included 'no hands up' and 'talking partners'. The teachers carefully planned a series of questions for the students to discuss with their pre-selected 'talking partner'. The students understood that their talking partners changed weekly and that they were selected randomly. Strategies 'no hands up' and 'talking partners' were simultaneously implemented to assist students' conversations and thinking skills. After having an opportunity to discuss

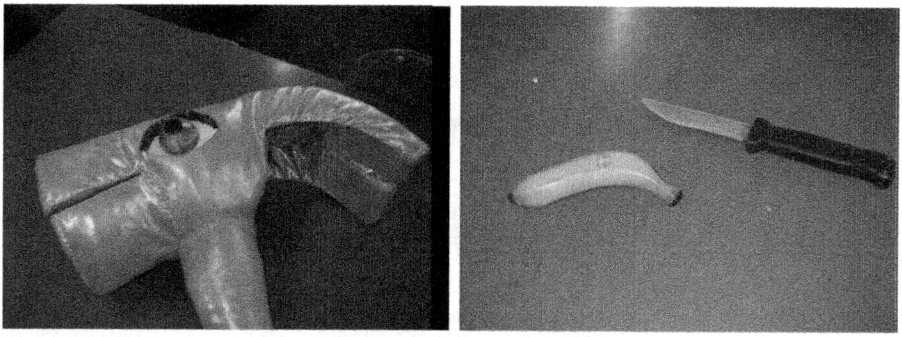

Fig. 8.6 A sample of props brought by the props manager of a local theatre company

their responses in pairs, Fleur and Clara randomly selected pairs to share their conversation ideas with the rest of the class. Students understood that any one person could be selected to respond on behalf of their pair. These questions followed targeted experiences and activities such as a visitor from a local theatre speaking about props and the showing of video of a stage play with clearly distinguishable props. In these cases teacher questions included:

1. Why are props important to a stage play?
2. If your bedroom was to become a scene for a play, which things would be the most important props? Why?
3. Tell us how you think props make plays better?
4. Of the props the theatre props manager showed us which one was the best and why?

   Imagine a play in which there is a scene with children having dinner in front of the television watching their favourite show.

5. What props would be needed?
6. What would they be made of?
7. Rank them according to their significance (Year 6 only).
8. What attributes would the props need to display if the play was being repeated for five consecutive nights? Justify the inclusion of each one (Year 6 only).

When facilitating these strategies, teachers must ask open-ended questions; students first discuss their ideas and think with their 'talking partner' before sharing their ideas with others. Students do not raise their hands in response to the teacher questions; rather the teacher randomly selects several 'pairs' to share their views.

Other strategies included the bringing in of community experts who can assist the students to make authentic connections to the real world. As mentioned above, in this study the props manager from a local theatre company visited both classes, demonstrated and discussed the characteristics and function of props in a stage show. Figure 8.6 shows a number of the examples he brought with him.

Fleur also used an activity called true/false, in which she made a series of statements that the students discussed in their talking partners, one at a time. Each pair

Fig. 8.7 Three of the final props designed and developed by students, the first two by Year 6 and the third by Year 2

considered whether the statement was true or false, with reason. The statements included the following: props must be big; a thimble is a good prop; and a banana cannot be a prop. After each pair discussed their ideas with each other, they then shared their discussions with the class when requested by Fleur.

At Year 6 Clara also used true/false and another activity called PCQ. In this activity the students discussed the pros and cons and identified questions to critique existing props in the first instance and later their own designs. One child in his final interview stated that this activity 'helped us think about what we needed to do to make our props'.

Engaging students in activities such as those outlined above facilitated their evaluation and synthesis of ideas to new situations. This is particularly useful in technology as when students are designing technological outcomes so they can draw from a range of experiences and knowledges to enhance their design ideas, process and the quality of their technological outcomes. Structured activities can result in a real change in learning. Evidence of learning, presented from this study, suggests this to be the case. The success of the final props, some of which can be seen in Fig. 8.7, was clearly evident at the final production.

Knowledge Perspective

The deployment conversation theme contributed to students' learning through a knowledge perspective. At times teachers explicitly drew on knowledge they knew the students already have; at other times students automatically deployed knowledge they had and which they understood would contribute their team's project. The fishing example used earlier illustrated this. Another group member suggesting that an oval is a good description of a fish shape recalling her learning in mathematics. In Clara's Year 6 class, the students were able to deploy measuring skills to design a scale model. They also brought knowledge from their parents' occupation to their technological practice, such as working with specific materials, wood and plastic, for example, or using bracing to join two sections of wood.

Funds of Knowledge

Students deployed knowledge and skills from their home and community, known as funds of knowledge, to assist and contribute to their learning in technology. This was the knowledge they already possessed and brought to their practice at times without specific prompting from teachers. Funds of knowledge influenced what students brought to their learning in technology. Student acquisition of knowledge, and then deployment of that knowledge into their technology project context, was a significant aspect of their learning. The analysis of the classroom talk indicated that students gained their knowledge for later transfer, from either their participation in activities with their families, interactions with artefacts or through social structures at home. I have called this use of funds of knowledge 'participatory enculturation'. This was illustrated by Alan and Dougal from Clara's class when they were selecting suitable materials for the stand of the microphone. Dowelling may come up as a possible option and a question was raised about how large (in diameter) it can be obtained. Dougal, whose father is a contractor, mentioned his dad had some quite large dowelling, which Alan then liken to a broom handle.

Dougal	My Dad had stuff about that big [indicates circle approximately 25 mm using the thumb and first finger].
Alan	Yeah broom handles are large dowelling.

Also to assist their technological practice, some students deployed knowledge from more passive activities such as watching TV or reading books. I have called this use of funds of knowledge 'passive observation'. This was illustrated when students recognised a waggon from a TV show and had knowledge about the 1930s microphones from watching old movies and their knowledge of fishing from reading a book.

When students brought knowledge from their home and culture to their technological practice, they were able to contribute to not only their group's technological outcome but their own and peers' technological context knowledge. By understanding the value of their own cultural practices, students put themselves in a better

Fig. 8.8 Mandy's autophotograph and comment about joining timber at 90°

position to assist their group, which in turn assisted the development of their self-esteem, a major contributing factor in students' learning. Students learned they had valuable contributions to make. The knowledge they took for granted as an integral part of their home and community culture was not known to their peers and subsequently they contributed significantly to their groups' technological practice.

An interesting finding from this study was that on a number of occasions when things got difficult for the students they drew on their funds of knowledge; such as when Mandy's group from Clara's class was attempting to join timber slats at 90 ° angles. Figure 8.8 shows Mandy's autophotograph of their challenging task. The text is the conversation Maddy and I had during her final stimulated recall interview about the photograph.

It is also illustrated by Rex's attempt to assist Debby and Issy to work cooperatively by suggesting they adopt his Dad's strategy for cooperating. 'What I used to do is if you there was two and there was one, so I did this, because my dad always says, "which one" and then the other two wanted two and then if there's one person who likes it, then we, we don't like it though' (interpretation-taking turns).

These findings have implications for teachers because they demonstrate that students learn from each other, and they all have home and community experiences that may contribute to others' learning. Teachers need to understand that students bring knowledge gained at home and in their community to technology education and use it to assist them in understanding and contributing when developing technological outcomes in a collaborative manner. It is therefore useful if teachers know their students and have an understanding of their cultural knowledges, skills and customs in order to assist deployment and enhance students' learning, social skills and self-esteem. With this understanding teachers are in a better position to plan units of work within authentically situated contexts and have the potential to motivate students by enhancing opportunities for them to implement cultural practices and knowledge from their homes and communities, to assist their own and others' learning. In order to be able to do this, teachers must first have knowledge of their students' cultural backgrounds and practices.

School-Based Learning Knowledge Transfer

While undertaking technological development, students also deployed knowledge from a range of other sources, to assist and contribute to their technological practice. They use knowledge learned in other learning areas and previous technology activities to assist their understanding of technological practice, for example, measuring and geometry mentioned above. As a technology unit progresses, students also deployed knowledge learned earlier within the current technology unit. Again imagine Clara and Fleur's classes. Early in the unit, they learned about props and their desirable attributes such as that props need to be durable and lasting, safe to use, easily identifiable, able to be seen and be in keeping with the historical and geographical location of the play. This knowledge was then be deployed later in the unit when the students designed their own outcomes.

This study found that the Year 6 students particularly implemented ideas they had gained from other school subjects without specific instruction to do so. Year 2 students also did this but less often. This counters research by Moreland and Jones (2000) and my own experience that students need to be specially taught to transfer knowledge from one learning area to another. One possible explanation is that the students in this study were highly motivated to develop quality outcomes, because they would be open to scrutiny from all those attending the school productions. Increased motivation to complete quality outcomes meant that students searched for ways of doing things well and therefore drew on knowledge and skills they had on hand as well as undertaking research where necessary. Teachers need to be cognisant of the impact an authentic context has on students' motivation to achieve in technology.

An implication of this for teachers is that students come to their technology projects with significant knowledge from their home, cultural and school communities. This knowledge comes in a range of forms and types and includes not only direct content knowledge but also process knowledge and knowledge about ways to behave and strategies for working collaboratively. Understanding this breadth of knowledge will enable teachers to prompt students' deployment of existing knowledge and skills through questioning and direct statements.

How This Might Be Used to Improve Teaching and Learning

The three most significant ways that teachers can use the findings from this study to improve teaching and learning are utilising funds of knowledge, planning and undertaking intercognitive conversations with their students to gain insight into students' learning in technology and planning opportunities for intercognitive conversations between students.

Teacher awareness that students' funds of knowledge can contribute significantly to their technological practice means that teachers can plan and implement opportunities for their students to be cognisant of and explore their own and others' relevant cultural knowledges related to their current project. The intercognitive conversation is a useful tool in this process. By using a range of activities and strategies such as the ones outlined in this chapter, teachers are able to engage with their students in a fashion that enhances the students' undertaking of their technological practice and will also assist teachers in developing deeper understanding of how students learn in technology, thus engaging them in 'teaching as inquiry' as outlined in the Effective Pedagogy section of The New Zealand Curriculum (Ministry of Education 2007, p. 35).

Further to this by facilitating the undertaking of student-to-student intercognitive conversation, teachers shift the focus of learning from the teacher to the students, thus embracing a student-centred approach to learning. To successfully implement student-to-student intercognitive conversations, teachers must teach specific strategies and attitudes to ensure students are open to others' ideas and flexible with their own thinking. Once achieved and intercognitive conversations are a natural part of the classroom culture, students are in a sound position to drive their own learning and assist their peers in theirs.

What Might Be Investigated Further

Although a number of potential investigations were identified as a result of this research, in this section I mention the five most relevant. The participants in the study worked collaboratively to design and construct their intended outcomes. In order to develop a single outcome as a group of three, collaboration and cooperation were essential. This study highlights the difficulties students had when working collaboratively. When working collaboratively students were forced to use intercognitive conversation with their peers in order to reach common understandings when different ideas were put forward. I believe there is potential for further study into students' ability to work collaboratively on a single project, while implementing a number of conversation strategies to assist in the collaborative process while protecting self-esteem. This would be particularly relevant to senior secondary schools where students are less likely to engage in collaborative technological practice despite it being commonplace in industry-based technological practice.

In this study students needed a range of knowledge and skills to assist their outcome development and construction process. The study identified two new sources of funds of knowledge. The first being participatory enculturation, in which students brought knowledge gained through active engagement in activities such as building tree houses with a father or fishing with a grandfather, and passive observation, in which knowledge gain came through noninteractive observation, such as watching movies and television. There is therefore potential to investigate these sources of funds of knowledge, to establish further insight into each and to determine the effectiveness of each and whether these are the only two sources of funds of knowledge, or are there others not identified through this research.

This study found that students in early primary school were able to develop mock-up designs to evaluate and modify their design ideas using a limited number of attributes to guide them. The Year 2 participants were able to articulate what a mock-up was and why it was made. This contradicts earlier findings and therefore opens opportunities for further research into how young primary school students' evaluate their technological outcomes using intermediate outcomes and attributes.

The study also found that students in Year 6 evaluated their outcomes using a greater number and more complex attributes than students in Year 2. In Year 6, students considered all their identified attributes: easily recognisable, durable, safe, era specific and ergonomically designed. In Year 2 only two of the five attributes were considered by any one participant in their product evaluations. This field has the potential for further investigation given the qualitative nature of this research and the small number of participants. It would be interesting to investigate whether it is a typical difference or whether it was unique to this study. Investigation could also be completed in related areas, such as the number and complexity of attributes able to be used by students at various levels of primary school.

In Year 6, the students understood the influences materials had on the quality and function of their final product. For example, Mandy understood that the wood would make a good frame for their radio but that it needed to be joined carefully. The radio group also realised that plastic coreflute could cover the frame and be painted to assist their design's authentic appearance. In Year 2, the students were not given an option of selecting suitable materials for their final outcome. Their teacher determined that the flying fish would be made from papier-mâché. This poses the question about the age and stage at which students are able to select appropriate materials for their outcomes to benefit the quality of outcomes and increase its likelihood of success, thus offering an implication for a researcher with the potential of a new field of investigation.

How Teachers Might Contribute to These Investigations

For teachers to be able to access the learning from this study, I recommend that they get their students working collaboratively on authentic technology projects. I suggest they facilitate students' movement out into the community to identify a real need with real clients and major stakeholders. By working collaboratively and embracing learning from a range of cultures and disciplines and working within an authentic context, students are able to extend their capabilities, knowledge and skills through the deployment of funds of knowledge. When engaging students in conversations, which facilitate synthesis, analysis and evaluation of materials and information, teachers are able to gain valuable insight into students' development of technological knowledge and concepts. In order to teach technology effectively, teachers need to have a good understanding of what students learn in technology and how that learning occurs (Jones and Moreland 2001). Also important is that by understanding the sources of deployed knowledge, teachers are in a better position to assist student deployment of this knowledge. Therefore, in classrooms where conversations about learning are a commonplace and a constructive environment is prevalent, then cognitive development is more likely to take place. Through conversation with their students and through listening to conversations among students, teachers are also able to gain insights into particular students' cognitive understanding in technology.

Teachers could further contribute to the above investigations by being aware of the importance of and undertaking the co-construction of outcome attributes as a guide to assist peer and self-product evaluation. They could also be purposeful in explaining to students how undertaking a range of modelling processes increases the likelihood of the development of successful outcomes.

Conclusion

Teachers should be encouraged to share their successes in technology education with their peers. Technological practice is usually a collaborative and cooperative activity. Classrooms are increasingly so, as teaching and learning pedagogies and practice align with learning in the information and digital ages and beyond. Successful teaching strategies and approaches could be shared on subject association websites such as TENZ (Technology Education New Zealand) and DATA (Design and Technology Association, UK). Subject association conferences are also another forum for sharing successful classroom practice. Such conferences frequently offer opportunity for teacher workshops in which ideas can be shared and discussed. The Ministry of Education in New Zealand also offers another medium (technology online) for sharing successful technological practice. I believe the current shift in many countries to collaborative teaching and innovative learning practices will assist this process and therefore assist researchers in technology education in the identification of potential research projects and participants.

An electronic copy of this PhD thesis can be found at this URL: http://researchcommons.waikato.ac.nz/handle/10289/7787

Acknowledgement I could not have undertaken this study without the unfailing love and support of my husband Mike Riley and the support, advice and direction of my three Chief Supervisors: Judy Moreland, Kathrin Otrel-Cass and John Williams. Thank you all for your time, input and patience. My second supervisor, Alister Jones, was my constant in this study. He saw value in my work when I saw only a room full of cotton wool. He made a significant contribution, not only to my thesis but also to my life.

References

Bloom, B. (Ed.). (1956). *Taxonomy of educational objectives: The classification of educational goals* (1st ed., Vol. 1). New York: Longmans, Green and Co Inc.

Jones, A., & Moreland, J. (2001). *Frameworks and cognitive tools for enhancing practicing teachers' pedagogical content knowledge*. Paper presented at the ASERA, Sydney.

Lichtman, M. (2006). *Qualitative research in education: A user's guide*. London: Sage publications.

Ministry of Education. (2007). *The New Zealand curriculum*. Wellington: Learning Media.

Moreland, J., & Jones, A. (2000). Emerging assessment practices in an emergent curriculum: Implications for technology. *International Journal of Technology and Design Education, 10*(3), 283–305.

Taylor, J. T., & Bogdan, R. (1998). *Introduction to qualitative research methods: A guidebook and resource* (3rd ed.). New York: Wiley.

Wolcott, H. F. (1988). Ethnographic research in education. In R. Jaeger (Ed.), *Complementary methods for research in education*. Washington, DC: AERA.

Chapter 9
Teaching Bioethics: The Intersection of Values and the Applications that Advances in Technology Make Possible

Deborah Stevens

Abstract This chapter summarises research that shows that in teaching bioethics, student-centred, narrative- and discussion-based pedagogy was critical to achieving specified outcomes. These outcomes included development of technological and scientific conceptual understanding, values appreciation, critical thinking, philosophical argument and improved skills of relating to others, managing self and participating and contributing. Students achieved these outcomes regardless of their academic history.

*As cultural, social, environmental and technological changes transform the world, the demands placed on learners and education systems are changing (*Australian Curriculum, Assessment and Reporting Authority, 2014).

Introduction

Bioethics education reflects that technology is a value-laden activity and seeks to prepare citizens to understand the technical processes, to assess the ethical issues and to actively participate in the decisions that are made with respect to the development and control of technology. Bioethics combines scientific and technical knowledge with knowledge of human value systems, including cultural and spiritual values. As developments and applications in science and technology affect the lives of individuals and the composition and evolution of society in unique ways, it is pertinent to ask how prepared current and future citizens are to make decisions with respect to unprecedented ethical and legal dilemmas. What values will govern the applications and possibilities that advances in technology make possible?

D. Stevens (✉)
University of Waikato, Hamilton, New Zealand
e-mail: debstevensmail@gmail.com

PJ. Williams, D. Barlex (eds.), *Contemporary Research in Technology Education*, Contemporary Issues in Technology Education, DOI 10.1007/978-981-10-2819-9_9

Inclusion of cultural and moral values and the development of key competencies have been at the forefront of international curriculum reform independently but simultaneously undertaken in New Zealand, Australia, the UK, the US and numerous OECD countries. For example, the New Zealand curriculum (hereafter NZC, Ministry of Education 2007) identifies five key competencies necessary for students 'to live, learn, work and contribute as active members of our communities' (p. 12) and 'who can make ethical decisions and act on them' (p. 10). The five key competencies are thinking. managing self, relating to others, participating and contributing and using language, symbols and texts. These competencies parallel those of international curricula including the Australian capabilities of critical and creative thinking, personal and social capability, ethical understanding and intercultural understanding.

The Questions I Asked and Why I Think They Are Important

In the 6 years prior to the research, I had written, revised and presented a stand-alone bioethics course to 800+ Year 11 to Year 13 students at an independent girl's school. My practical experience was that the teaching of bioethics as a dedicated subject was able to address the identified need to develop scientific literacy within students, while simultaneously teaching and developing the competencies and values required by the curriculum. Facilitation in a high socio-economic independent school was atypical, however, and whether such a curriculum would be a valuable tool for developing technological and scientific literacy and the teaching of values and the key competencies in all schools (state, integrated and independent), from all socio-economic levels required investigation. It was this path that led me to pose the following questions for research:

- Affective outcomes: In what ways does the teaching and learning of bioethics as a stand-alone subject contribute to the development of a participating student's personal values, moral reasoning and worldview?
- Cognitive outcomes: In what ways does the teaching and learning of bioethics as a stand-alone subject contribute to the development of a participating student's cognition, including academic learning and critical thinking?
- Curriculum outcomes: How do the affective and cognitive outcomes demonstrated by students participating in the bioethics curriculum relate to the values and key competencies requirements of the NZC (Ministry of Education 2007)?
- Pedagogical framework: In what ways does the student-centred, narrative- and discussion-based pedagogy encourage students to explore their values and develop judgement including the use of argumentation and evidence-based reasoning? And what are the wider implications for curriculum delivery of this pedagogical framework?

How I Tried to Answer the Questions

I sought a school that reflected New Zealand's ethnic and socio-economic diversity, in which to conduct case study research. The 1200+ student body at Koru College (a pseudonym) is around 66 % New Zealand European, 20 % Maori, 4 % Pacific Islander and 10 % other, including Asian and African, and international students from South America and Europe.

Following my approach to the Principal, I was invited to facilitate a 90-min full staff meeting. Ten teachers from across a variety of academic disciplines expressed interest in participating in the year-long research. Following separate presentations to Year 11 and Year 12–13 students that gave a sample of the bioethics curriculum, there was significant interest in self-selecting to participate in the bioethics project. In the end practical constraints of timetabling identified two teachers and 78 participating students. The 22 students from Year 11 were taught by Nick, a fifth year teacher, whose primary teaching discipline was English. Including a group of ten accelerate students, the Year 11 student group, who gave up their only timetabled study period to participate, was skewed to the upper end of the academic bell curve.

The 56 Year 12 and 13 students were taught by Helen, an educator with some 25 years' experience who had been the head of the transition department for the past 5 years. While the full range of academic abilities was represented by students in the transition department, this student cohort also included students from the Athena Unit, an on-campus supported learning environment for students with physical and intellectual impairments. Accordingly, the academic bell curve in the Year 12–13 group was skewed to the lower end of the range. As the bioethics course was timetabled during one of the students' four transition periods of the week, the Year 12–13 students who self-selected to participate in bioethics committed to completing their transition syllabus in three rather than four periods per week.

Helen, Nick and I worked together for several days during the summer holidays to select the content for the bioethics project from my previously developed curriculum. Helen and Nick found the case of baby Theresa, an anencephalic baby whose parents wished to donate her organs to be harvested and transferred to infants on a transplant waiting list, compelling and the research curriculum developed from this. Grounded as it is in organ donation, something most senior secondary students have thought about as they are required to tick a box declaring them an organ donor or not when they apply for their driver's licence, Theresa's story was seen as a starting point that had some relevance for participating students. From this scientific starting point of transfer of living tissue between members of the same species (allotransplantation), the ethical theories of utilitarianism, Kantian ethics, natural law, virtue ethics and situation ethics were explored. Figure 9.1 expands on how these concepts were taught and linked.

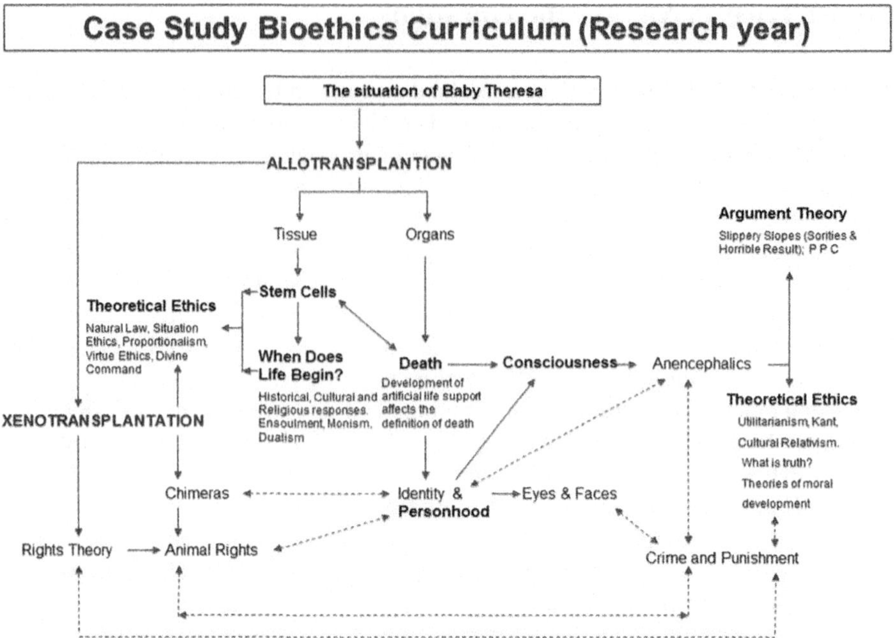

Fig. 9.1 The case study curriculum (*solid arrows* indicate lesson sequence and *dashed arrows* indicate how these topics linked in and were referred back to)

In summary, the curriculum delivered by both teachers to the two case study groups covered:

1. Scientific and technological concepts, for example, tissue transfer between members of the same and different species, development of artificial life support and development of human reproductive technologies
2. Philosophical concepts, for example, argument theory, specific ethical theories, when life begins and the concept on personhood

Timetabled for 1 hour each week for the full academic year, the curriculum was taught using narrative, where stories, wherever possible authentic, were used for student capture and theoretical instruction and also as practical activities for student experience and learning. Throughout the course, narratives were experienced through an extensive variety of formats, media, and genres.

Mixed qualitative and quantitative data was obtained throughout the year-long project. Students anonymously completed two Likert-style surveys: a brief initial survey administered one-third of the way through the course and a comprehensive end-of-course (EOC) survey. Pre- and post-teaching and learning activities, including a pre- and post-teaching survey to investigate a student's understanding of scientific terms and the ethical, cultural and spiritual concerns surrounding stem cell research, were performed to assess the learning of philosophical and scientific concepts.

Semi-structured interviews (averaging 22 min each) were conducted mid-course and EOC with 40 key student informants (KSI) and at the beginning and end of the year with the two collaborating teachers and the school principal. In addition to observing every lesson and keeping a researcher journal, every lesson was MP3 recorded. Lesson plans, collaborating teacher journals, informal debriefs and minutes from group meetings with collaborating teachers completed the qualitative component of the research.

The range of data sources allowed for a systematic, cross-examination (triangulation) approach to the data analysis. Division of the student cohort into two case study groups (Year 11, and Years 12–13), and replication of the bioethics curriculum in two contrasting settings within the same general school environment, afforded additional comparison of the data generated from the mixed sources, increasing the reliability and validity of the outcomes. Descriptive and statistical analysis was undertaken with the range of data collected, including the use of Excel and SPSS computer programmes, to apply a variety of statistical tests and the creation of a valid and reliable construct for *affective and cognitive response to bioethics at Koru College.*

What I Found Out

Three interrelated findings emerged:

1. The outcomes observed for participants in this research arose from the narrative-stimulated, discussion- and activity-based pedagogy utilised in the bioethics curriculum.
2. The participating students had both an affective and a cognitive response to the bioethics curriculum regardless of their position on the academic spectrum.
3. The bioethics curriculum proved an effective vehicle for comprehensive theoretical–cognitive and character–behavioural values education, the development of key competencies and the gaining and retention of technical, scientific and philosophical knowledge.

As described, the bioethics curriculum trialled in this research was anchored by exploration of applied ethical issues where narratives were used as the stimulus material for student-led discussion. Narratives were often presented using videos, songs and static images conveyed using PowerPoint and Smart Board, often being accessed directly through the Internet. Students perceived these technologies as instantly recognisable, relevant and appropriate to their personal lives and learning. Movie clips, videos, music, graffiti art, billboards and creative use of activity-based media such as dramas, role plays and games proved engaging modes of narrative through which bioethics was taught and learnt. Both collaborating teachers presented the inherently controversial bioethical dilemmas in ways that aroused students' curiosity and incentivised engagement. This included creating suspense and leaving the situation tantalisingly unresolved for a period within, and occasionally between, lessons.

Participating students unanimously reported that narratives concerning real people and situations made the curriculum content interesting, relevant and meaningful to them. Topics covered were 'like stuff we should know' and like 'no other subject'. Narratives presented curriculum content in ways that assisted students to identify the bioethical issue and the perspectives of the central characters and incentivised understanding of the development and application of the technology. Students' capacities to imagine and be empathetic to life stories and worldviews other than their own were enhanced, together with their understanding of the values that underpin a variety of cultural, social and religious perspectives. All students developed their personal competencies of self-awareness and self-critique, listening carefully, following complex plots and observing and identifying patterns of meaning and argument in both their own and others' communication. Students acknowledged that learning to consider different perspectives assisted them to discipline their initial tendency to instantly dismiss as wrong or misguided the views of others that were in conflict with their own. Students learnt to understand that others may respond differently and to recognise what these different perspectives may be and the values that anchor them. Students progressively learnt to take turns, not to interrupt and to listen attentively as bioethical scenarios were inquired into throughout the year. Further, critical thought that led to the construction of a cogent argument capable of withstanding the analytical scrutiny of others resulted in the scrutinising of intuitive 'gut reactions', and students reported that they learnt not to respond emotionally 'off the top of their head'. Classroom interaction developed relationships between the students and enabled practice in the competencies of managing self, relating to others and participating and contributing.

Significantly, the student focused, narrative- and discussion-based pedagogy integral to the trialled curriculum resulted in participating students spontaneously reporting that they felt like cocreators of knowledge, rather than passive receivers of standardised information. Rather than being told what was right or what was required by the teacher, the majority of students reported that the teaching method used in the bioethics course necessitated and permitted them to think actively. Through practical thinking and active participation in discussion by listening and contributing, students perceived themselves to be more directly and interactively involved in the creation of their knowledge within the bioethics class. Dan provides an example:

> In bioethics, we kind of, we are being really original, rather than doing 'learning' of what's already been done. We are coming up with our own ideas and then working on them and discussing them, rather than doing something, or learning something that has already been decided on. Whereas in another subject you are just being told what you need to do and what you need to understand and you just work away at understanding it. (Dan, Year 11, 100625–07)

Every student who participated in the course reported improved thinking skills in their anonymous EOC written survey. This was reflected in the interviews with each KSI. The majority of responses incorporated the concept of thinking 'more' including thinking about more perspectives, exploring 'more ideas' in their thinking, thinking 'more widely', being 'more questioning' in their thinking, thinking more deeply, taking more time to think, listening more carefully and thinking through the consequences of different choices.

> It makes you think—like *think*. It's not like when you go into class, you get your book out, and you just write it down ready for an exam. It's not like that. When you go in there you look forward to doing it and you wonder what you're going to learn about. And it's stuff that you can kind of take with you. Like you keep it in your mind; you keep it in your head, but then it's like something you can always talk about with someone else. It's important because it's real-life stuff. (Holly, Year 12, 101101–03)

A theme of the bioethics class as 'open' emerged during KSI interviews. Related to the theme of students creating their own reasoned views, the learning field was perceived as more level, in that generally a number of ethical approaches to a dilemma could be taken. KSIs reported this led to an appreciably reduced sense of 'being wrong', which in turn boosted individual student's confidence and sense of autonomy and empowerment. Students all heard the same scenarios, but were free to explore, challenge and respond in their own way. This involved distinguishing their personal response, explaining their response and supporting, defending and/or refining their response. In this way, the content within the bioethics course was perceived as more open ended, more open to personal interpretation and opinion and more open to the sharing of these. In survey and interview, the majority of participating students (95 %) perceived the opportunity provided in the bioethics class to engage with their personal values, to debate relevant real-world issues and to learn about the values of others as something not generally available to them in their current schooling.

The 'open', student-focused teaching and learning within the bioethics course induced a positive effect within students including a subtle experience of feeling good in the class, of feeling confident in their learning and confident and satisfied that their contribution and participation were pertinent, appropriate and of value. The experience of positive good feelings at the end of a lesson was variously described as 'exhilarated', 'energised' and 'that I've learnt something new'. Students reported feeling empowered and experiencing a sense of autonomy and satisfaction through the choice-making, narrative-based activities and through having to advocate for their personal view. This sense of autonomy was boosted further by experiencing self-direction in the regulation of personal behaviour, including the decision to contribute, and their response to the differing views and opposing arguments of others. Participation in the narrative- and discussion-based activities enhanced participating students' sense of competence and, therefore, their perceived student identity.

Students' emotional and cognitive engagement in a lesson was physically represented through behavioural engagement including paying attention, leaning forward, tracking class interaction and active listening. Once engaged, emotional, social and academic learning proceeded within these diverse learners. Figure 9.2 and the explanation beneath it summarise the ways the narrative- and discussion-based teaching method engaged participating students across both case studies.

It may sound as if the positive cognitive and affective outcomes for students in the two diverse case study groups within this research investigation are too good to be true. The case study approach 'provides a unique example of real people in real situations' (Cohen et al. 2007, p. 253) and affords the researcher the opportunity to employ a broad assortment of quantitative and qualitative data collection techniques

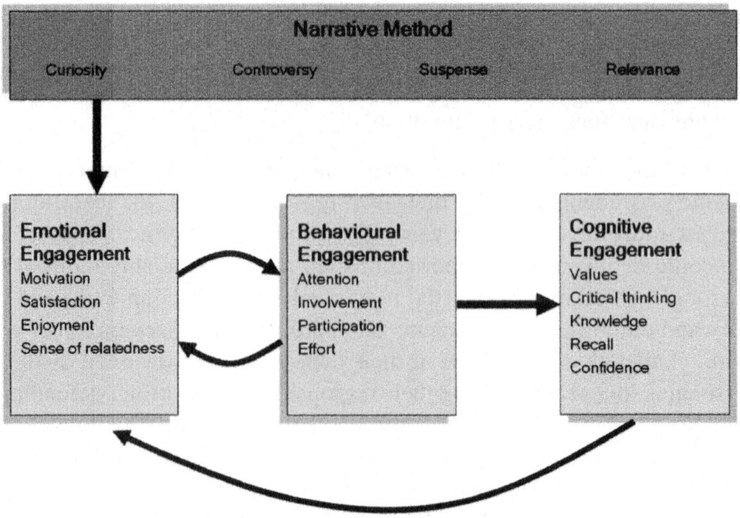

Contemporary ethical issues that were perceived by students as relevant to their current and future lives were presented through narrative using culturally relevant media. The use of authentic narratives in a variety of genre aroused curiosity, gained students' attention and incentivised emotional, behavioural and cognitive engagement through the use of controversy and suspense, while simultaneously humanising the bioethical dilemmas, and enabling the development of empathy. Discussion of the narratives enhanced students' experiential engagement with the characters within a dilemma, created interaction, connection and relatedness between classmates and a sense of being co-creators, rather than passive recipients, of knowledge. Induced positive emotions, including enjoyment, satisfaction, and confidence enhanced behavioural engagement, empowering individual students to higher levels of involvement, participation and effort. The rewards and satisfaction embedded in participation in the class increased students' intrinsic motivation to learn. Narratives were easily recalled and the academic knowledge that they anchored was retained.

Fig. 9.2 How the use of narrative in the trialled bioethics curriculum promoted emotional, behavioural and cognitive engagement

and, therefore, to apply a variety of analytical procedures. However, case studies are contextually bound; that is, it is not intended that the findings and deductions be generalised beyond the context in which a case study occurs (Cohen et al. 2007; Hammersley and Atkinson 2007; Yin 2009). Research to assess the generalisability of these research findings across schools from a range of socio-economic and demographic settings within New Zealand and a wider international setting is required. However, it is observed that the affective, cognitive and social outcomes demonstrated for students participating in this teaching bioethics project parallel the 'behaviour, life view and value view' (Jinhua 2008, p. 77) and 'science and technology', 'society' and 'family' (Wang 2008, p. 73) outcomes expressed by teachers and students who participated in a 14-week discrete bioethics course offered at the High School Affiliated to Beijing Normal University.

Recognising the affective and cognitive outcomes for participants in the research, Koru College maintained a stand-alone bioethics course in the timetable in the year following the trial, expanding the time allocation from one hour to two hours per week. This had curriculum planning implications. If bioethics went into the timetable, what was to come out? The remainder of this chapter describes the experience of the collaborating teachers with respect to curriculum planning, lesson planning, classroom management and assessment with the aim of sharing ideas useful for improving teaching and learning in the area of technological and scientific literacy.

How this Might Be Used to Improve Teaching and Learning

Developments in technology and the acquisition of new knowledge stimulate questions about the nature of the human condition, the nature of a good life and how humans ought to live. These are deeply philosophical questions. While an individual does not have to be an academic philosopher to think critically and concisely about ethical issues, the skills and knowledge required to teach the sciences differ from those required to teach ethics, something teachers themselves acknowledge (Grace 2006; Levinson 2001, 2003, 2004). Recognising the need for a different type of knowledge, many science and technology educators feel inappropriately qualified and under-resourced to be addressing ethical issues within their lessons (Hall 1998; Jones 2007; Levinson 2001; Levinson and Turner 2001; Macer et al. 1996). Science teachers expressed feeling pedagogically challenged with respect to the discussion- and activity-based teaching methods including storytelling, drama and role play, useful in the teaching and learning of ethics. Further, that there may be no clear solution to an ethical issue is challenging to the concrete descriptions and explanations frequently inherent in teaching science and technology.

(i) **Resources**

The collaborating teachers each had some, though limited, experience of ethics—Nick, philosophy, and Helen, business ethics. Prior to and throughout the research year, Helen and Nick worked closely with me, with each other and independently to gain knowledge, to plan lessons and to collect and make resources for teaching bioethics. Each of these activities was time consuming. Nick and Helen worked collaboratively sharing resources they produced including games, PowerPoint, props and scripts for short dramas, together with resources they discovered including Internet links, video and movie clips, books and articles. Such collaboration reduced the individual time commitment required of each teacher. Sharing what they felt worked and did not work improved lesson quality as each new topic was introduced. In this way a Koru College bank of bioethics resources, which has been drawn on and built upon following the research year, was begun. A list of international resources to support secondary school bioethics teaching is provided at the end of this chapter. Notwithstanding, the experience of the collaborating teachers within this research illustrates the need for resources for the teaching and learning of bioethics and for the need for teacher professional development and collaboration.

(ii) **Classroom management**

Bioethics is an applied subject, and as such, it requires a practical or experiential component, including provision of opportunities for students to learn ethical strategies and to argue rationally as they endeavour to reach an ethical conclusion (De Luca 2010; Jones et al. 2007: Levinson 2003, 2006; McKim 2010; Reiss 1999, 2003, 2010; Ryan 2008; Saunders 2009). The use of narrative and discussion proved fundamental to the positive affective and cognitive outcomes experienced by students in this research. However, the efficacy of a narrative- and discussion-based pedagogy to engage and teach students has particular implications for classroom practice, as the following quote from Nick illustrates:

> Certainly it was overwhelmingly exciting to see everyone so wanting to talk on this issue, or this point—'I want to respond to what she just said!' and 'Woop!—OH!—I ...!' People bursting with excitement at this idea of learning; at this idea of new things to them. Topics they had not come across but which they soon became very passionate about. (Nick, 101118)

Strategies were required to establish orderly student-led (rather than teacher-directed) philosophical 'argument' of controversial topics, and these took time to embed.

Hand and Levinson (2012) and Tomas (2010) observe that discussion of controversial issues is appreciably improved when students are equipped with skills to analyse and evaluate arguments. After establishing what philosophical arguing was not (with the aid of Monty Python), both teachers facilitated a lesson on how to 'argue' philosophically, including identifying underlying premises to an argument, checking to see if the conclusion followed logically from the premises and learning to anticipate counterarguments, at the start of the course. These elements were referred back to and modelled as the course progressed. Students assimilated these strategies over time.

One example from the early Year 11 case study illustrates how Nick skilfully employed silence to establish the student-led teaching method, including students 'generating' knowledge for themselves, the practice of patience, taking turns and allowing every student, not just the talkative ones, an opportunity to contribute. Nick had described the real-life case of baby Theresa. Central participants within this case disagree, their opposing views clearly articulating respective consequentialist (utilitarian) and deontological (Kantian) philosophies. Having described only the case, its participants and their opposing views, and without mentioning either philosophical school of thought by name, or discussing any of their underlying principles, Nick provided the class with a series of laminated cards including terms such as 'judge', 'parents', 'means to an end', 'greater good', 'going to die anyway' and

Fig. 9.3 Nick builds on student-generated learning at the completion of the silent card sorting activity

more. Showing the students the cards labelled 'utilitarian' and 'Kantian', Nick then instructed the class to use the cards to make two columns on the floor to:

> Work out what these terms might mean, and what the foundational beliefs behind each of these moral ways of thinking is. The activity will be made more interesting as you will complete it in silence. If you think the placement of a card is wrong, move it. Look through them and see if you think they follow the same argument. (Nick, classroom MP3, 100317)

Nick then stepped back as the students began to sort the cards with intensity, humour, stealth and at times almost 'charade-like' non-verbal communication.

As the cards were progressively sorted with 'baby Theresa' being left in the middle by silent agreement, Nick asked if all students were 'happy with the arrangement?' Agreement was nodded. Nick was then able to build on this learning and begin teaching the academic concepts and terms associated with each of the ethical theories (Fig. 9.3). However, this was not before he had the student's themselves identify more through asking them 'What would someone who is a utilitarian think about moral issues?' The activity took just over 15 min and included a detailed exploration and definition of deontological and utilitarian philosophies, including an introduction to the Kantian concept of universal rules and categorical imperatives.

Engaging students in learning, fostering classroom interaction and building relationships with and between students were important aspects of the narrative-stimulated, discussion-based teaching method used in the research. Narrative/stories appeared not simply to affect the individual but also the classroom environment, helping to create an ambience of safety, equality and respect and enhancing affective engagement. Essential to this were the experiential elements and social interaction that occur as stories are shared. However, the use of narrative raises certain questions for the individual classroom practitioner. These questions include what might constitute narratives that are inclusive of the diverse cultural and ethnic groups within society, from where these narratives might be sourced and how the use of such culturally inclusive narratives in a curriculum may be maximised. Facilitation of inquiry into and discussion of ethical issues and the values and beliefs that underpin the plurality of responses to them requires impartiality. Narratives used within a bioethics curriculum are generally told in particular contexts for particular purposes and are therefore constructed and presented in certain ways. The question of how the personal beliefs and biases held by the facilitating teacher may be minimised so that student exploration and autonomy are maximised is therefore raised.

(iii) **Assessment**

Ninety-five per cent of participating students reported that the bioethics class was different from other classes. The absence of board and bookwork and the dominance of discussion, together with the openness of participation this afforded, were identified as essential differences. Students were not required to take notes and there were no compulsory formal assessments during the year. However, pre- and post-teaching and learning activities, in the form of digital and written surveys, were performed to assess the learning of philosophical and scientific concepts. Throughout the year, learning was evidenced through the development and correct use of academic vocabulary, the appropriate application of technological and philosophical concepts and through developed skills of argument and communication. Comparison of MP3 recording of lessons over time enabled the analysis of development in communication skills, including the use of appropriate terms and language, analytical skills and critical thinking. The practical, active thinking and discussion aspect of the course was referred to by every KSI as they described how they could 'take away' and remember the content from the bioethics course.

Helen did develop an internally assessed level three, unit standard, which if passed contributed four credits towards a student's National Certificate of Educational Achievement (New Zealand's national secondary school qualification). Completion of this standard was not compulsory, but voluntarily. Both Nick and Helen noted how compulsory assessment might change the way the trialled bioethics curriculum was taught. However willing a teacher may be to diversify teaching methods and place students at the centre of learning, the constraints of an assessment- and standard-driven curriculum may mean that didactic, hierarchical teaching methods, with an emphasis on how to pass assessments, are retained for reasons of efficiency.

Through his participation in the trial, Nick recognised that the focus of his teaching in his principal discipline of English was primarily on teaching skills associated with assessment, for example, the structure required by a well-crafted essay, rather than with the subject content knowledge. Following this realisation, Nick began to alter his teaching practice in English to include greater inquiry through the exploration of values and ethics-related questions, encouraging students to use their imaginations to explore the worldviews of characters within the narratives (novel, film and poetry) being studied in English. Above all, following his experience in the bioethics class, Nick encouraged students in his other classes to talk and discuss:

> I think that is the main thing I have learnt; just let the students talk. Don't cut off, 'Alright, we're moving on. No. We're moving on'. Let them have their say, because that is where a lot of the offshoots, those beautiful little stems, came from; that just letting students talk. (Nick, 101118)

Nick completed the English syllabus with his students in the research year. This would indicate that formal assessment and narrative- and discussion-based pedagogy need not be adversative.

What Might Be Investigated Further

If, as research participants indicate, formal assessment inhibits the use of creative, practical, student-focused activities and influences the scope of learning by encouraging focus on how an assessment response is structured rather than subject knowledge, the question of how values and competency outcomes can be assessed becomes more complex. Questions implied within the area of values and competency assessment include how responses that are not judged right or wrong, but well supported or not supported, are to be assessed; how in a time-constrained assessment- and outcome-focused education environment teachers may be encouraged to provide more space for student discussion; and if this space is provided, how teachers are to incorporate acknowledgement of sound thinking. These questions together with exploration of whether every aspect of education needs to be assessed require further research and policy changes.

How Teachers Might Contribute to These Investigations

As Helen observes, outcomes with respect to competencies and ethical decision making may not be immediately visible:

> In terms of measurability of results right now, I think that some of the things we have done in bioethics might not have results for a few years. I think it is sitting there—the seeds are there and the ideas are there—and I think the students will come up with ideas [] and they will think 'Oh—I remember when we…. (Helen, 101118)

It would be good to test the anecdotal evidence that teachers feel restricted in their ability to incorporate practical, student-centred, narrative- and discussion-based pedagogy into their classrooms, through further research. Formal research would give teachers an opportunity to contribute by vocalising their positions while generating evidential data. In the absence of a formal study, teachers may contribute to changing a standardised, outcome-driven approach to education by progressively incorporating practical, narrative- and student-centred teaching activities as Nick did with his English syllabus. From here teachers could share their professional knowledge and student-based evidence with policy-makers and curriculum developers within their schools and then at regional and national levels.

Technology is a value-laden activity where ideas are generated, assessed and selected on perceived worth (Forret 1997; Jones 2007). Developments in science and technology determine the features of the world citizens inhabit now and in the future. If students are going to think deeply about issues and engage in deliberation with respect to developments in science and technology and their impacts on society, law, commerce, culture and politics, bioethics education needs to be incorporated into the curriculum in ways that are resonant and relevant to the students and rigorous in academic approach. I hope this chapter goes some way to providing useful insights and information.

International Resources to Support Secondary School Bioethics Teaching

- BioethicsBytes offers multimedia resources for the teaching of bioethics, available at http://bioethicsbytes.workpress.com
- The Nuffield Council on Bioethics http://www.nuffieldbioethics.org
- and http://blog.nuffieldbioethics.org/
- and http://www.nuffieldfoundation.org/science-society.
- The Wellcome Trust charity (http://www.wellcome.ac.uk) resources provided in a number of genre including book and online and PD courses for teachers.
- The Biotechnology Learning Hub including http://www.biotechlearn.org.nz/themes/bioethics/frameworks_for_ethical_analysis.
- BEEP http://www.beep.ac.uk/. The site is rich with case studies on a wide variety of scientifically based bioethical issues. Online discussion forums for both teachers and students are also provided through the site.
- The University of Iowa Bioethics Outreach Programme http://www.bioethics.iastate.edu provides a large section containing case studies and hypotheticals for use in the classroom and professional development courses (both in-house and online) for teachers.
- GENIE, available at http://www.le.ac.uk/ge/genie/
- techNyouvids http://www.youtube.com/user/techNyouvids
- A video-based resource that aims to raise awareness about emerging technologies. This site includes videos that explore critical thinking and logic, ethical issues raised by technologies and the history of science and information on how various technologies work (e.g. transistors, nanotechnology, synthetic biology).
- An assortment of multilingual resources for educators may be found at http://www.unescobkk.org/rushsap/resources/shs-resources/ethics-resources/multilingual-material/ including a Bioethics Core Curriculum. Available in many

different languages, an English version may be downloaded from http://unesdoc.unesco.org/images/0016/001636/163613e.pdf
- A series of books produced by the UNESCO including the *Asia-Pacific Perspectives on the Ethics of Science and Technology* (2007), *Perspectives on Bioethics Education* (2008) and *Perspectives on Biotechnology* (2008) present papers on the teaching and learning of bioethics. While the majority of these pertain to the tertiary sector, they provide a useful professional development for teachers, in addition to some of the examples being adaptable to the secondary classroom situation. These books may be downloaded through http://www.unescobkk.org/rushsap/resources/shs-resources/ethics-resources/. This site also provides a variety of other ethics resources and links.
- The Young Foresight Project: A UK initiative in design creativity involving mentors from industry, in B France and V Compton (Ed) *Bringing Communities Together: Connecting Learners with Scientists or Technologists*. Rotterdam: Sense. Develops a 'designing without making' approach that seeks student responses to questions such as 'what would you do with this technology?' or 'what would you like to see scientists do with this technology?'

An electronic copy of this PhD thesis can be found at this url: http://researchcommons.waikato.ac.nz/handle/10289/7746

Bibliography

Jones, A., McKim, A., & Reiss, M. (2010). Towards introducing ethical thinking in the classroom: Beyond rhetoric. In A. Jones, A. McKim, & M. Reiss (Eds.), *Ethics in the science and technology classroom* (pp. 1–5). Rotterdam: Sense Publications.

Macer, D. (2008). *Moral games for teaching bioethics*. Haifa: United Nations Educational, Scientific and Cultural Organizations Chair in Bioethics Office.

Willmott, C., & Willis, D. (2008). The increasing significance of ethics in the bioscience curriculum. *Journal of Biological Education, 42*(3), 99–102.

Zeidler, D. L., Sadler, T. D., Applebaum, S., & Callahan, B. E. (2009). Advancing reflective judgement through socioscientific issues. *Journal of Research in Science Teaching, 46*(1), 74–101.

References

Australian Curriculum, Assessment and Reporting Authority. (2014). *The Australian curriculum*. http://www.australiancurriculum.edu.au/generalcapabilities/ethical-understanding/introduction/introduction. Last accessed 20 Oct 2014.

Cohen, L., Manion, L., & Morrison, K. (2007). *Research methods in education* (6th ed.). London: Routledge Press.

De Luca, R. (2010). Using narrative for ethical thinking. In A. Jones, A. McKim, & M. Reiss (Eds.), *Ethics in the science and technology classroom* (pp. 87–101). Rotterdam: Sense Publications.

Forret, A. (1997). *Learning electronics: An accessible introduction*. Unpublished doctoral thesis, University of Waikato, Hamilton.

Grace, M. (2006). Teaching citizenship through science: Socio-scientific issues as an important component of citizenship. *Prospero, 12*(3), 42–53.

Hall, E. (1998). Science education and social responsibility. *School Science Review, 81*(295), 14–16.

Hammersley, M., & Atkinson, P. (2007). *Ethnography: Principles in practice* (3rd ed.). London: Routledge Press.

Hand, M., & Levinson, R. (2012). Discussing controversial issues in the classroom. *Educational Philosophy and Theory, 44*(6), 614–629.

Jinhua, F. (2008). The actualities and prospect of bioethics education in a Chinese middle school. In D. Calderbank & D. Macer (Eds.), *Asia Pacific perspectives on bioethics education* (pp. 77–79). Bangkok: United Nations Educational, Scientific and Cultural Organization.

Jones, A. (2007). Reflecting on the development of a new school subject: The development of technology education in New Zealand. *Waikato Journal of Education, 13*, 273–294.

Jones, A., McKim, A., Reiss, M., Ryan, B., Buntting, C., Saunders, K., … Conner, L. (2007). *Research and development of classroom-based resources for bioethics education in New Zealand*. Hamilton: Wilf Malcolm Institute of Educational Research.

Kopelman, L. (2006). Bioethics as a second-order discipline: Who is not a bioethicist? *Journal of Medicine & Philosophy, 31*(6), 601–628.

Levinson, R. (2001). Should controversial issues in science be taught through the humanities? *School Science Review, 82*(300), 97–102.

Levinson, R. (2003). Teaching bioethics to young people. In R. Levinson & M. Reiss (Eds.), *Key issues in bioethics* (pp. 25–38). London: RoutledgeFalmer.

Levinson, R. (2004). Teaching bioethics in science: Crossing a bridge too far? *Canadian Journal of Science, Mathematics, and Technology Education, 4*(3), 353–370.

Levinson, R. (2006). Towards a theoretical framework for teaching controversial socio-scientific issues. *International Journal of Science Education, 28*(10), 1201–1224.

Levinson, R., & Turner, S. (2001). *Valuable lessons*. London: Wellcome Trust.

Macer, D., Asada, Y., Tsuzuki, M., Akiyama, S., & Macer, N. Y. (1996). *Bioethics in high schools in Australia, Japan & New Zealand*. Christchurch: Eubios Ethics Institute.

McKim, A. (2010). Bioethics education. In A. Jones, A. McKim, & M. Reiss (Eds.), *Ethics in the science and technology classroom* (pp. 19–36). Rotterdam: Sense Publications.

Ministry of Education. (2007). *The New Zealand curriculum for English-medium teaching and learning in years 1–13*. Wellington: Learning Media Ltd.

Reiss, M. (1999). Teaching ethics in science. *Studies in Science Education, 34*, 115–140.

Reiss, M. (2003). How we reach ethical conclusions. In R. Levinson & M. Reiss (Eds.), *Key issues in bioethics* (pp. 14–23). London: Routledge Press.

Reiss, M. (2010). Ethical thinking. In A. Jones, A. McKim, & M. Reiss (Eds.), *Ethics in the science and technology classroom* (pp. 7–17). Rotterdam: Sense Publications.

Ryan, B. (2008). *Ethics education in New Zealand: A literature review (directed study)*. Hamilton: University of Waikato Centre for Science and Technology Education Research.

Saunders, K. J. (2009). *Engaging with controversial science issues: A professional learning programme for secondary science teacher in New Zealand*. Doctoral thesis, Curtin University of Technology, Western Australia.

Tomas, L. (2010). *Merging fact with fiction: A study of Year 9 students' developing scientific literacy through the writing of hybridised scientific narratives on a socio scientific issue*. Doctoral thesis, School of Mathematics, Science & Technology Education, Queensland University of Technology, Australia.

Wang, L. (2008). This year's flowers are redder than last year: A brief introduction of the bioethics project in the high school affiliated to Beijing Normal University in 2003–2005. In D. Chalderbank & D. Macer (Eds.), *Asia Pacific perspectives on bioethics education* (pp. 73–76). Bangkok: United Nations Educational, Scientific and Cultural Organization.

Yin, R. K. (2009). *Case study research: Design and methods* (4th ed.). Thousand Oaks: Sage.

Chapter 10
Exploring the Role of Professional Learning Communities in Supporting the Identity Transition of Beginning Design and Technology Teachers

Denise MacGregor

Abstract Wenger (Communities of practice: learning, meaning and identity. Cambridge University Press, Cambridge, 1998) argues that we cannot become humans by ourselves. Similarly, this chapter purports that we cannot become teachers by ourselves. The process of becoming a teacher is complex and is influenced by many occurrences, practices and people (Pillen M, Beijaard D, Brok PD, Eur J Teach Educ 36:1–21, 2012). When pre-service teachers commence and complete their university study, they hold varied personal narratives about their perceived professional identity as educators (Lortie D, Schoolteacher: a sociological study. University of Chicago Press, Chicago, 1975; Groundwater-Smith S, Mitchell JM, Mockler N, Learning in the middle years: more than a transition. Thomson Learning, South Melbourne, 2007; Smith R, Teach Teach Theory Pract 13(4):377–397, 2007). These perceptions have been shaped by many constructs including personal and professional histories (Day C, Kington A, Stobart G, Sammons P, Br Educ Res J 32(4):601–616, 2006; Furlong C, Eur J Teach Educ 36(1):68–83, 2013), the content of university courses (Zuga K, J Technol Educ 2(2):60–72, 1991; Smith R, Teach Teach Theory Pract 13(4):377–397, 2007), school- and community-based experiences and interactions and conversations with peers, university and school staff, friends and family.

Once pre-service teachers transition into their teaching roles, these perceptions are influenced further by the diversity and complexities of the school context itself (MacLure M, Br Educ Res J 19(4):311–322, 1993; Coldron J, Smith R, J Curric Stud 31(6):711–726, 1999). This chapter explores the impact of schools as professional learning communities and more specifically the role that school-based mentors and nontraditional mentoring strategies can play in supporting beginning teachers. The chapter argues that while the role of school-based mentor teachers is significant, it is the establishment of a whole school culture built on a positive

D. MacGregor (✉)
University of South Australia, Adelaide, SA, Australia
e-mail: denise.macgregor@unisa.edu.au

P.J. Williams, D. Barlex (eds.), *Contemporary Research in Technology Education*, Contemporary Issues in Technology Education, DOI 10.1007/978-981-10-2819-9_10

organisational climate that is instrumental in providing support. The chapter concludes with a discussion of traditional and nontraditional strategies that can enable schools to establish such cultures.

The Questions I Asked and Why I Think They Are Important

The main research question is: How can schools successfully support the transition of beginning teachers into the profession?

It is argued that few experiences in life have such a tremendous impact on the personal and professional life of a teacher as does the first year of teaching (Feiman-Nemser 2003; Findlay 2006). It is during the first year of teaching that: *[t]he initial experiences are imprinted, embedding perceptions and behaviours regarding teaching, students, the school environment, and their role as a teacher* (Findlay 2006, p. 511). These authors suggest that the first year of teaching should be seen as a phase in learning to teach and that beginning teachers should be greatly supported by school leadership, colleagues and the wider school community throughout this time. Furthermore, existing research has concluded that schools have an obligation to provide a supportive environment through which beginning teachers can further develop their professional knowledge and an understanding of their role as a teacher (see, e.g. Tickle 2000; Flores 2001; Carter and Francis 2010; Ingersoll and Strong 2011). Throughout Australia, Europe and the United Kingdom, increased centralised requirements have placed schools in a more prominent and influential role with regard to both pre-service and in-service teacher professional development. This in turn has led to the establishment and implementation of a diverse range of school-based mentoring and induction programmes in schools.

Carter and Francis (2010, p. 250) describe mentoring as a process that: *[m]itigates teacher isolation, promotes the concept of an educative workplace and that leads to the creation or understanding of consensual norms in schools or faculty.* More specifically, research has found that mentoring is used to address issues of teacher survival, skill development and, ultimately, retention in the profession. In Australia, 25–40 % of beginning teachers leave the profession in their first 3–5 years of teaching (Berliner 2001; Ewing and Smith 2003). These figures are drawn from beginning teachers who have secured permanent employment. A lack of adequate support from school administrators and colleagues has been identified as a significant contributing factor in beginning teachers' decisions to leave the profession (Ingersoll and Strong 2011).

Harrison et al. (2006) and Carter and Francis (2010) have argued strongly that those who undertake the role of mentor have the potential to play a significant role in assisting beginning teachers in not only developing their professional knowledge but also their professional identity. However, defining the role of the mentor is complex and definitions are diverse. For example, Clutterbuck (1992, p. 71) states that: *[a] mentor is a more experienced individual, willing to share his/her knowledge with someone else less experienced in a relationship of mutual trust.* Clutterbuck's definition places the emphasis on the notion of coaching, or the passing on of knowledge, to facilitate the development of specific skills or capabilities.

The literature also warns that there is the possibility that a poorly organised mentoring process or the acquisition of a mentor who is not an appropriate role model can constrain the learning and confidence of beginning teachers (Ballantyne et al. 1995). More specifically in the field of Design and Technology education, beginning teachers may find themselves assigned to a mentor who is limited in their knowledge of changing technologies such as advanced manufacturing and digital technologies or in contemporary constructivist approaches to learning. As Ballantyne et al. (1995, p. 302) argues:

> The danger exists that mentors contribute to the professional socialisation of beginning teachers into the use of traditional techniques, rather than assisting the carry-over of progressive techniques into the classroom.

The consequence is that the mentor may become antagonistic towards the beginning teacher and in so doing tension arises. In most instances, when this tension occurs, the beginning teacher, who is in a probationary situation, complies, and the status quo of the school context is maintained, leaving the beginning teacher feeling disillusioned and disempowered.

The literature clearly identifies that the role of the mentor is multidimensional and complex and can be highly influential in supporting the transition of beginning teachers into the profession. The study that informs this chapter sought to explore both the complexity and effectiveness of the mentoring role in a Design and Technology teaching context.

How I Tried to Answer the Question

The study that informed this chapter adopted an interpretive, qualitative research method approach combining narrative inquiry and case study methods. The study of narrative is the study of the ways in which humans experience the world. Creswell elaborates further and argues that narrative design *involves drawing portraits of individuals and documenting their voices and their visions within a social and cultural context* (Creswell 2008, p. 513). It is argued that narrative is especially suited to conveying the complexities of the classroom, the nature of teachers' knowledge and the development of professional identity. For this reason, the adoption of a narrative methodology was deemed the most relevant and appropriate to achieve valid and insightful findings for this study. More specifically, the use of a narrative methodology enabled the collection and retelling of the stories of the beginning teachers with a high level of authenticity.

The type of data analysis adopted for this study was narrative analysis (Yin 2003). As a distinct form of qualitative research, narrative analysis focuses on the study of the individual or a small group of participants. Data were gathered through the collection of stories that reflected the individual teacher's experiences, and data were analysed through the retelling of the teachers' stories, that is, the narratives of teachers were retold by the researcher. Described as *a unique qualitative analytic procedure used only in narrative research* (Creswell 2002, p. 52), this analysis provided the researcher with a deeper insight into the experiences of the individual. For example, in the following section of this chapter, one of the beginning teachers re-storied narrative of his professional relationship with his mentor providing insights into the emotional processes associated with beginning to teach. Thus, the data analysis was both a description of the story and themes that emerged from it.

The participants for the study were a group of five first year Secondary Design and Technology teachers who had completed a 4-year undergraduate teaching programme specialising in Design and Technology Education. The settings for this study were the five schools in which each of the participants had commenced teaching. Data were collected over a 15-month period using a questionnaire that included open-ended text response questions, a teacher professional knowledge framework, three semi-structured interviews and reflective e-journal entries.

What I Found Out

The Role of School-Based Mentoring in Shaping Beginning Teachers' Professional Identity

Findings from the five case studies concurred with the views espoused in the literature, that is, that mentoring is a key strategy for the effective induction and support of beginning teachers (Feiman-Nemser 2003; Carter and Francis 2010; Ingersoll and Strong 2011). Mentors played a significant role in supporting the majority of the beginning teachers to successfully transfer into school-based settings. However, the data revealed that the degree of formality in both assigning and assuming the mentoring role varied considerably between school sites. For example, only two of the five beginning teachers had been officially assigned a mentor. In one case, a teacher or a number of teachers within the faculty or school voluntarily assumed the mentoring role. In two cases, beginning teachers were not mentored at all.

The findings of this study revealed two major mentoring roles that impacted on shaping beginning teachers' professional identity and on their role as beginning teachers. The mentoring roles included providing:

- Personal and emotional support
- Teaching-related assistance and advice, including feedback on professional practice

Participants identified personal and emotional support as being one of the most valuable roles of the mentor. Participants valued a mentor who was approachable and someone who they were able to talk to, to seek advice from and, at times, confide in about their anxieties and limitations without fear of being judged. The retold narratives that follow highlight the significant role that school-based mentor/s and a supportive school culture played in shaping teachers' professional knowledge and identity in the early stages of transition. The narratives commence with Jason's story.

Jason was initially assigned as a mentor who was also the faculty coordinator. Jason's narrative, as captured in interview 1, identified the role of the mentor teacher as being instrumental in building his confidence, in supporting his personal well-being and in shaping his professional identity. In contrast, the second part of this narrative, as told in the second interview, reflects Jason's angst when his mentor became ill.

The Value of a School-Based Mentor: 'My Mentor Really Cared About Me and My Teaching' (Jason)

Jason commenced teaching with a 1-year contract at a metropolitan Catholic boys' college and became a permanent member of staff in his second year of teaching. In his first year of teaching, the school was undergoing a change of leadership, with the appointment of a new principal and deputy. Jason noted at the time that:

> There is a lot of change everywhere within the school ... I think change is good and I think it has been easier for me to come into a school that is transitioning. It means that there is change for everybody and not just for me.

Jason acknowledged that his confidence as a teacher improved significantly throughout the first term, mainly through the support of his assigned mentor teacher, who was also the coordinator for Design and Technology. Jason stated:

> My mentor teacher has been there to support me in everything. I would just ask and he knew the answers, he is the guiding hand.

However, in the second term, Jason's mentor teacher took leave as a result of illness. During this time, Jason recalled that:

> I didn't have anyone to fall back on, I felt lost at first. I didn't realise how much I had come to rely on him at both a professional and personal level ... I did get a bit more guidance as the term progressed ... but it was never the same. The level of support was different. My mentor really cared about me and my teaching. He enabled me to get through my first term of teaching.

When Jason's mentor returned later in the year, the professional and personal support recommenced.

For Jason, the first 6 weeks of teaching were very emotional; it was a time of uncertainty and of questioning his own teaching abilities, particularly in subjects with which he was not familiar. It was during this time that the role of the mentor became crucial in providing empathy and reassurance. The centrality of the role of the mentor in supporting beginning teachers throughout this emotional period is clearly evident in Jason's narrative.

Transition Made Easy: 'Smooth as Silk' (Brenton)

In Brenton's retold narrative, the value of being immersed in a professional environment in which mentoring relationships with a number of teachers are developed is highlighted. For the majority of beginning teachers in this study, and as exemplified by Brenton, mentors were instrumental in providing teaching-related assistance and advice. Evidence from the data also suggested that mentoring relationships were strongest when beginning teachers and the mentor/s taught the same subjects, topics or year levels. According to Brenton, the school-based initiatives and being mentored by colleagues from the same faculty made the transition into teaching:

> As smooth as silk; all staff offered as much help as I needed. I was buddied up with a teacher from the Design and Technology faculty and with a Year 8 home group teacher. The Design and Technology staff encouraged me to take risks and to make mistakes from which I could learn. All the teachers in the school have been really supportive and welcoming, smiling, greeting me, asking about me, how I am going … this has really made a big difference about how I feel when I am at the school.

An effective mentoring programme, coupled with welcoming staff, enabled Brenton to develop increasing confidence in his role as a teacher as the year progressed. Although Brenton was assigned a specific mentor, he found that all staff within the Design and Technology faculty and throughout the school assumed responsibility in supporting his transition. The staff had established a positive organisational climate (Carter and Francis 2010) in response to their recognised belief in the benefits of mentoring. In Brenton's school, there was an established ethos of supporting beginning teachers. Brenton further stated that:

> I was supported to reflect or self-critique. I was encouraged to sit in on other teachers' classes if I wanted to. I was given several repeat classes so this gave me an opportunity to reflect and fine-tune any aspects of the lesson. This provided me with a foundation [from which] to establish the expectations for each class. (Brenton, e-journal entry, Week 6)

Evidence from this study suggested that working in close proximity with a mentor/s who taught the same subjects or classes also enabled informal discussions and conversations to take place throughout lunch and recess breaks. It appeared that the day-to-day pressures of classroom teaching meant that finding the opportunity to set time aside for more formal or regular meetings was difficult.

Whole School Induction: 'Time to Be Prepared' (Travis)

The value of whole school commitment to the support of beginning teachers was also evident for Travis. While Travis continued to be voluntarily mentored by colleagues throughout the year, including members of the school leadership team, he believed that the 1-week, school-based induction for all new staff at the beginning of the year provided him with the opportunity to be successful in his teaching. Whole school induction provided Travis with the opportunity to understand the school infrastructure and to be introduced to and welcomed into the school culture.

Travis commented that:

> This allowed me to have seven days of professional preparation with my colleagues. It enabled me to get a hold of all of the awkward things like getting to know other teachers' names, know where rooms are, understand behaviour management policies, in fact everything to do with the physical teaching environment.

As a result Travis stated that:

> I was able to feel very comfortable when the students arrived. I think this is a fantastic reflection on the management of the school. (Travis, in-school stage, e-journal, entry 1)

For the two participants who were not mentored, the transition into teaching was described as a time of isolation and uncertainty. The retold narratives that follow articulate Aaron's and Damien's initial shock and struggle to survive.

Isolation: Going It Alone (Aaron)

Aaron commenced his permanent teaching appointment in a rural area school. He was the only Design and Technology teacher on staff. He replaced the former Design and Technology teacher who had taught in the school for over 30 years. The facilities in which he was to teach had not been updated throughout this time. Aaron stated:

> I was just handed the keys and pointed in the direction of the Design and Technology workshops. I am unsure if this is the same for all new teachers at the school or if it's simply because of the learning area that I am in and that I am by myself.

For Aaron, being the only Design and Technology teacher on the school staff was initially a source of challenge and anxiety. From the commencement of his teaching position, he was expected to assume full responsibility for all administration, maintenance and teaching in the Design and Technology faculty. For example, Aaron was expected to order and prepare budgets for materials and equipment, to maintain equipment and to oversee occupational health, safety and welfare (OHSW) as well as to plan, teach and assess student learning.

While initially viewing this responsibility as a source of tension, Aaron was able to look forward and predict that the experience was a learning opportunity. Aaron attributed the support of his wife and members of the wider community as being instrumental to his survival in the first months of teaching. He also identified his maturity and level of personal resilience, developed through past work experiences, as enabling him to adapt to the roles and responsibilities he now needed to assume.

Like Aaron, Damien was not supported by a mentor. His first e-journal entry revealed initial self-doubts, as he asked himself: 'Can I do this?' Initial e-journal entries stated that little support was offered by colleagues in the first few weeks: 'a time when I really needed it the most'. Damien further explained that:

> The other D&T teacher is also new to the school so we have been left to set things up ourselves. There were no programs in place when we started; what I teach and how I teach it has been pretty much up to me. As a result, there were times when I thought 'I can't do this'.

The beginning teachers in this study made continued reference to how they believed they were being perceived and ultimately accepted by their mentor/s and by the more experienced colleagues in the faculty. Working in practically based settings with a range of diverse materials and equipment meant that the beginning teacher's level of competence very quickly became apparent to colleagues. This study identified that once mentors and colleagues deemed that they could competently and safely perform their teaching role and that it aligned to some degree with their own practices, the beginning teachers became accepted and were then encouraged to take increasing ownership in regard to lesson content and how they taught. The tensions identified by Ballantyne et al. (1995) in being assigned a mentor who could constrain the learning and confidence of beginning teachers were not evident in this study. In fact, the beginning teachers in this study were able, after a relatively short period of time, to demonstrate a capacity to exercise independence, to introduce new content knowledge into the learning area and, in some instances, to adopt innovative pedagogical approaches.

Feiman-Nemser argue that beginning to teach revolves around several themes including *reality shock and the lonely struggle to survive* (Feiman-Nemser 2003, p. 26). These authors argue further that the organisation of many secondary schools does not support collaboration between teachers, thus reinforcing a *sense of isolation and autonomy* for the beginning teacher (Feiman-Nemser 2003, p. 26). An analysis of Aaron's and Damien's narrative supports this argument. However, for the other participants in this study, *the reality shock and lonely struggle to survive* (Feiman-Nemser 2003, p. 26) appeared to be mitigated through effective induction or mentoring. Beginning to teach was viewed by all teachers in the study as a time of uncertainty and, for a number of them, a time of self-doubt.

The personal and emotional support provided by mentor/s and/or an induction process appeared to be pivotal in providing support throughout this time. The role of mentors in providing teaching-related assistance and advice as well as opportunities for critical reflection and feedback on professional practice appeared to be instrumental in supporting this process. The literature (see, e.g. Tickle 2000; Flores 2001; Carter and Francis 2010; Ingersoll and Strong 2011) states that mentoring is used to address issues of teacher survival, skill development and, ultimately, retention in the profession. Carter and Francis (2010) argue further that the most effective mentoring processes emerge from a positive organisational climate in schools. It is the establishment of such a climate in the form of a professional learning community that is explored in the following section of this chapter.

How This Might Be Used to Improve Teaching and Learning

The literature and findings of this study suggest that the first year of teaching provides beginning teachers with an immersed experience of participation within varied professional learning communities (Wenger 1998). Engagement with members of that community, acquiring increasing competence in it, taking on some of the perspectives and aligning oneself within that community appear to positively support transition and shape professional identity. While the narratives presented earlier reflect the transition of beginning Design and Technology teachers, a case can be made for the establishment of professional learning communities to support all new entrants into the teaching profession. It can be argued (Lovett and Cameron 2011) that without support, beginning teachers will change schools or leave the profession. As a consequence, much depends on the ability of a school as a whole and for teachers individually to develop a culture that is able to provide ongoing support and to sustain beginning teacher's professional learning. This section of the chapter provides suggestions on how schools generally and how experienced teachers can establish professional learning communities to support the personal and professional growth and well-being of their beginning teachers. Strategies that will be addressed build on traditional mentoring strategies to include the use of social media, providing reduced teaching contact time to enable regular engagement in professional development opportunities, establishing formal partnerships with professional teaching associations and providing beginning teachers with shared teaching and learning spaces. The role and importance of school leadership in providing support mechanisms for beginning teachers are also discussed.

Mentoring and E-Mentoring

Traditional methods of mentoring include the development of a one-on-one professional relationship between an experienced teacher (mentor) and a beginning teacher (mentee). Through this relationship, the mentee gains both content and pedagogical knowledge, confidence and emotional support. Simultaneously, the mentor is provided with an opportunity to reflect on their own practice. The process of mentoring and being mentored can be viewed as a learning opportunity for both parties. While the findings of this study found that the most effective aspects of the mentoring role were through providing personal and emotional support and professional advice, it can be argued that similar degrees of effectiveness could be achieved through social media and more specifically through e-mentoring.

When researching the impact of mentoring in an organisational context, Single and Single (2005) suggest that all the benefits associated with traditional mentoring can be realised through e-mentoring. According to their findings, the impartiality that was derived from communicating through an online environment enabled a high level of honesty and openness to develop. Single and Single (2005) argue further that this degree of openness would have been more difficult to achieve if only meeting in a face-to-face situation. In the study that is central to this chapter, participants communicated with the researcher through e-journals. E-journal entries were used to follow up interviews and promote further conversation. E-journals in the form of reflective emails were used by participants to add thoughts or ideas that occurred to them between interviews. On several occasions, the researcher initiated a follow-up email to all participants, while, on other occasions, participants would email to inform the researcher of a situation, an idea or just for the purpose of an informal chat. As stated by Single and Single (2005), the informal nature of the e-journal facilitated open and frank online conversation, where questions could be asked, discussions would ensue and solutions could be derived.

In schools the combination of traditional and nontraditional mentoring strategies could be effectively utilised. Initial one-on-one mentoring relationships could be enhanced through email, or other social media such as Facebook, to overcome the time constraints imposed on both the mentor and mentee through daily teaching commitments. In England the Facebook site *Product Design Surgery* enables newly qualified teachers and recently qualified teachers to share ideas (https://www.facebook.com/groups/productdesignsurgery/).

For beginning teachers who commence teaching in rural and remote settings, the opportunity to be e-mentored by a more experienced colleague in a neighbouring school could reduce the sense of isolation and the reality shock of beginning to teach (Feiman-Nemser 2003). Online professional development opportunities can also provide beginning teachers with the opportunity talk with other beginning teachers who may be undergoing similar situations and experiencing the same emotions.

Social media, including e-mentoring, can be easily and frequently accessed and allows individuals to form groups online which offer the opportunity to share information and collaborate with colleagues in geographically diverse settings. The merging of traditional mentoring strategies with social media provides an additional dimension for professional learning communities to support beginning and experienced teachers.

Professional Development

In the states and territories of Australia, school-based induction programmes are complemented by a number of online professional development programmes established and funded through private providers and through the various Departments of Education. Web-based resources such as www.aussieeducator.org.au provide beginning teachers with a number of programmes, resources and tools (including 'survival guides'). Similar online resources are also available nationally, for example, in the United Kingdom through organisations such as the Teacher Development Trust http://TDTrust.org/. The aim of these professional development programmes is to offer beginning teachers with readily accessible, long-term sustainable support.

However, the effectiveness of such resources is dependent upon user awareness in knowing that such resources exist and on opportunities for regular accessibility. In Australia there have been continued calls for the need to reduce beginning teachers' contact hours to support the transition into full-time teaching. It can be argued that the first year of teaching should be viewed as a phase in learning to teach (Feiman-Nemser 2003; Findlay 2006). Beginning teachers need time to understand the complexities of their school, including administrative procedures and classroom management policies, and they need time to further develop their pedagogical and subject content knowledge. Providing beginning teachers with a reduction in contact hours will enable them to maximise learning not only from more experienced colleagues in the school site, through peer observations of practice and critical discussion, but through accessing and engaging regularly with online professional development programmes.

Long-term formal and informal induction programmes that extend beyond the first week or few weeks of term and are structured as a component of a reduced teaching load can demonstrate to beginning teachers what is needed to sustain their motivation, commitment and job satisfaction (Lovett and Cameron 2011). Literature (Lovett and Davey 2009; Reeves 2008) states further that beginning teacher's satisfaction and ultimately retention in the profession are more likely to remain high if opportunities to observe colleagues, to be mentored and to have access to professional development opportunities extend beyond the first year of teaching. Reeves (2008) attributes 40 % of beginning teacher's learning to professional development opportunities, including attending seminars and workshops and online professional development websites. The remaining 60 % is attributed to long-term interactions with site-based colleagues, through induction and mentoring, and to experiences with students and parents. Thus, establishing a whole school culture of support and collegial learning that extends beyond the first few weeks of teaching is vital to ensuring beginning teachers' learning and satisfaction and ultimately to their retention in the profession.

Professional Associations

The place of professional associations in providing a collegial space for support and learning for beginning teachers deserves greater recognition. It can be argued that if beginning teachers are to solve problems in their practice (Nieto 2003), the collaboration and sense of professional belonging that associations can provide are vital. Nieto (2003) suggests that the collaborative discourse generated through meeting with colleagues allows beginning and experienced teachers to build critical and long-standing relationships and that these relationships produce a sense of community. Professional associations provide such communities. In each of the states and territories of Australia, the Design and Technology Teachers' Association (DATTA) provides regular professional development opportunities. In South Australia DATTA-SA provides a series of workshops and seminars each school term. In addition, each year a representative from the South Australian association is invited by the University to meet with the final year Design and Technology teachers. Graduates are welcomed to the profession and encouraged, through free association membership, to attend professional development opportunities and to access the association website (www.datta.sa.edu.au/). Similarly in the United Kingdom, Teachmeets (http://teachmeet.pbworks.com/w/page/81578876/Teach%20Meet%20South%20Manchester%20July%202014) are an informal movement for beginning and more experienced teachers to share experiences. Attendance at professional development seminars can enhance teacher collegiality through providing an opportunity to talk about teaching and teaching practices. Attendance also provides an opportunity to introduce beginning teachers to new and innovative teaching practices in a supportive environment.

Working in Shared Learning Spaces

Many beginning teachers commence teaching in isolation in classrooms, often with the door closed to the outside world. As a consequence, limited opportunity is provided for informal observation of teaching practice and the resultant constructive feedback, or for the spontaneous conversation, through which frustrations can be voiced and successes shared. To maximise support and collaborative problem-solving, teachers need to have shared workspaces and tasks and common planning time. Shank (2005) argues that these factors are invaluable for helping beginning teachers to make sense of teaching.

The nature of teaching in Design and Technology education dictates that teachers generally work in shared spaces (workshops) where shared planning and collaborative problem-solving can occur. Unlike the views espoused in the literature, the beginning Design and Technology teachers who were participants in the study that is central to this chapter were supported by colleagues in both formal and informal ways. The beginning teachers, who taught in a faculty with other Design and Technology teachers, were provided with support in planning and developing workshop skills and techniques and with words of advice and encouragement.

As Brenton stated in the final interview of the study:

> What continues to shape my professional identity are the interactions with staff. This continues to build my self-confidence, for example, when staff give you praise or encouraging comments for the work you have done. There have been a lot of teachers who have done that this year so that I now feel accepted and welcomed into the community.

Shared workspaces and collaborative planning sessions enabled the development of professional relationships that were effective in supporting professional growth and development.

The Role of School Leadership

The literature (Fernandez 2000; Flores 2004) argues that there is widespread acceptance of the role that school leadership can play in developing school cultures that promote professional development for beginning teachers. A common trait of effective school leadership is the ability to build, promote and maintain a professional learning community through which teachers, including beginning teachers, are able to develop a sense of self-efficacy and self-worth (Fernandez 2000). Such leaders not only develop individualised professional relationships with beginning teachers, but they ensure that the infrastructure of the school provides support and professional development mechanisms on a number of levels. At the macro level, it is to work in collaborative ways to create common goals, a vision for the future and standards for the school. Lovett and Cameron (2011) state that school leadership and more specifically principals are instrumental in building *individual capacity for the collective good* (2011, p. 93). The establishment of such a school culture enables all teachers to take responsibility for supporting the initial transition and learning of beginning teachers. At a micro level, the role of the principal is to ensure that each teacher is made to feel that they are a valued member of the school community and to continue to support them, in collaboration with the school staff in becoming so. The impact of the principal in knowing and valuing members of staff is reflected in the comment from Travis (interview 2) which states:

> I was really touched when the principal knew it was my birthday and left a card on my desk. The card said I was doing a great job. It made me feel great!

What Might Be Investigated Further

In this section of the chapter, the author explores more broadly what else schools, including member of the leadership team and teachers, should know about mentoring to successfully support the transition of beginning teachers into the profession. Firstly, mentoring programmes need to be planned, structured and coordinated within a schools' organisational policy. Mentoring programmes should also be developed with a view to longevity, that is, longevity commencing with a whole school long-term commitment to providing high quality induction and mentoring programmes that are implemented from year to year, through to individual mentoring arrangements between mentees and mentors. Findings from the study that informed this chapter would suggest that mentoring is most successful if provided for at least 6 months and preferably for a year or longer. Evidence suggests that extended mentoring enables beginning teacher to move from seeking answers to purely functional or administrative questions to seeking advice to improve classroom teaching practices to improve student learning outcomes. As evidenced by Travis's comment after 1 year of teaching:

> My confidence has really grown over the last year. This is due to the people I work with, leadership, colleagues, through exploring what I can teach in Design and Technology, taking it further and doing bigger and greater things. The support of my colleagues and leadership has been really vital.

Additionally, to be successful, mentoring programmes need to be viewed as being mutually beneficial to the mentor, mentee and school community as a whole. To facilitate such a view, members of the leadership team need to assume responsibility in ensuring that the goals of the programme are clear and known to all members of staff, that professional development opportunities are provided to support the development of mentoring skills and attributes, that mentors and mentees are well matched and that release time from their normal teaching load is allocated to the mentor teacher to facilitate their role. Evidence suggests that mentoring relationships are strongest when the mentor/s and mentees teach the same subjects, topics or year levels. It appeared that working in close proximity with a mentor/s who teaches the same subjects or classes enables informal discussions and conversations to take place throughout lunch and recess breaks. The nature of the conversations was often driven by issues that were specific to the classroom context, for example, maintaining student safety when working with equipment. It appeared that the day-to-day pressures of classroom teaching meant that finding the opportunity to set time aside for more formal or regular meetings was difficult.

Whole school support and commitment to the mentoring policy also need to be evident. Surrounding beginning teachers with a professional culture (Ingersoll and Strong 2011) that supports their professional growth and well-being is a significant factor in shaping professional identity throughout the first year of teaching. There is a need for mentoring programmes to be funded and the outcomes of these programmes to be acknowledged and measured in terms of not only addressing the current low levels of beginning teacher retention rates but in providing opportunities for sharing ideas, knowledge and expertise. In this way, teachers, both experienced and those new to the profession, can continue to move the learning area of Design and Technology education forwards through innovative and collaboratively informed practice. Graduates who are applying for a teaching position should be asking about the schools' mentoring programme and seeking clarification about the mechanisms for developing collegiality within the school. Beginning teachers should recognise that they are interviewing the school as much as the school is interviewing them.

How Teachers Might Contribute to These Investigations

There is a need to critically assess current mentoring programmes to ask, 'What kinds of mentoring program exist in schools and are they helpful to beginning teachers?' These are fundamental questions that remain relatively unanswered. There appears to be limited research and inquiry into the range of mentoring programmes offered by schools and the effectiveness of these and online programmes in supporting beginning teachers to transition into their roles. Increased classroom-based inquiry into mentoring programmes could serve to provide school leaders, teachers and schools with a current assessment of mentoring programmes. In doing so, aspects of those programmes that are effective, sustainable and transferable could be implemented across school sites.

This study examined the influence of mentoring on the development of professional identity of Design and Technology graduates from a 4-year university degree. However, there are other, shorter pre-service programmes for Design and Technology teachers where less time can be given to providing the knowledge and skills necessary for teaching, and it would thus be useful to examine the effectiveness of mentoring programmes for graduates from these programmes. In light of the high attrition rates for beginning teachers, more extensive research needs to be conducted on how beginning teachers can be better supported during this time of transition beyond mentoring programmes.

Conclusion

This chapter was premised on the belief that professional learning communities play a central role in supporting the successful transition and retention of beginning teachers into the teaching profession. Analysis of the narratives from the study that underpins this chapter suggests that beginning teachers' professional well-being and identity are enhanced through effective mentoring and induction programmes coupled with effective school leadership. This chapter provided suggestions on how schools and in-service teachers can establish professional learning communities to support the personal and professional growth of their beginning teachers. These strategies included building on traditional mentoring strategies to include social media, providing reduced teaching contact time to enable regular engagement in professional development opportunities, establishing formal partnerships with professional teaching associations and providing beginning teachers with shared teaching and learning spaces. The chapter also calls for an assessment of current mentoring programmes with a view to identify aspects that are effective, sustainable and transferable to be highlighted and implemented across school sites.

In summarising, it is the responsibility of all members of a professional learning community to contribute to the development of mechanisms to support collegiality and to the professional growth and well-being of those within that community.

Acknowledgement I wish to sincerely thank my principal supervisor, Professor Howard Middleton, who guided my learning and supported me professionally throughout every stage of my study. I also wish to sincerely thank my two co-supervisors Dr Margaret Freund for the initial years of my study and Dr Judith Peters for the latter part of my study. They both provided valuable guidance, support and friendship.

References

An electronic copy of the thesis can be found at; https://www120.secure.griffith.edu.au/rch/items/a911e276-50ba-436f-9020-9f582f8a9db6/1/

Ballantyne, R., Hansford, B., & Packer, J. (1995). Mentoring beginning teachers: A qualitative analysis of process and outcomes. *Educational Review, 47*(3), 297–308.

Berliner, D. (2001). Learning about and learning from expert teachers. *International Journal of Educational Research, 35*(5), 463–482.

Carter, M., & Francis, R. (2010). Mentoring and beginning teachers' workplace learning. *Asia-Pacific Journal of Teacher Education, 20*(3), 249–262.

Clutterbuck, D. (1992). *Mentoring*. Henley: Henley Distance Learning.

Coldron, J., & Smith, R. (1999). Active location in teachers' construction of their professional identities. *Journal of Curriculum Studies, 31*(6), 711–726.

Creswell, J. W. (2002). *Educational research: Planning, conducting, and evaluating quantitative and qualitative research* (1st ed.). Upper Saddle River: Pearson Prentice Hall.

Creswell, J. W. (2008). *Educational research: Planning, conducting, and evaluating quantitative and qualitative research* (3rd ed.). Upper Saddle River: Pearson Prentice Hall.

Day, C., Kington, A., Stobart, G., & Sammons, P. (2006). The personal and professional selves of teachers: Stable and unstable identities. *British Educational Research Journal, 32*(4), 601–616.

Ewing, R., & Smith, D. (2003). Retaining quality teachers in the profession. *English Teaching: Practice and Critique, 2*(1), 93–113.

Feiman-Nemser, S. (2003). What new teachers need to learn. *Educational Leadership, 60*(8), 25–30.

Fernandez, F. (2000). Leadership in an era of change: Breaking down the barriers of the culture of teaching. In C. Day, A. Fernandez, T. Hauge, & J. Moller (Eds.), *The life and work of teachers. International perspectives in changing times* (pp. 239–255). London: Falmer Press.

Findlay, K. (2006). Context and learning factors in the development of teacher identity: A case study of newly qualified teachers during their induction year. *Journal of In Service Education, 32*(4), 511–532.

Flores, M. A. (2001). Person and context in becoming a new teacher. *Journal of Education for Teaching, 27*(2), 135–148.

Flores, M. A. (2004). The impact of school culture and leadership on new teachers' learning in the workplace. *Teaching and Teacher Education, 25*(1), 814–825.

Furlong, C. (2013). The teacher I wish to be: Exploring the influence of life histories on student teacher identities. *European Journal of Teacher Education, 36*(1), 68–83.

Groundwater-Smith, S., Mitchell, J. M., & Mockler, N. (2007). *Learning in the middle years: More than a transition*. South Melbourne: Thomson Learning.

Harrison, J., Dymoke, S., & Pell, T. (2006). Mentoring beginning teachers in secondary schools: An analysis of practice. *Teaching and Teacher Education, 22*(8), 1055–1067.

Ingersoll, R., & Strong, M. (2011). The impact of induction and mentoring programs for beginning teachers: A critical review of the research. *Review of Educational Research, 81*(2), 201–233.

Lortie, D. (1975). *Schoolteacher: A sociological study*. Chicago: University of Chicago Press.

Lovett, S., & Cameron, M. (2011). Schools as professional learning communities for early career teachers: How do early career teachers rate them? *Teacher Development: An International Journal of Teachers Professional Development, 15*(1), 87–104.

Lovett, S., & Davey, R. (2009). Being a secondary English teacher in New Zealand: Complex realities in the first 18 months. *Professional Development in Education, 35*, 547–566.

MacLure, M. (1993). Arguing for yourself: Identity as an organising principle in teachers' jobs and lives. *British Educational Research Journal, 19*(4), 311–322.

Nieto, S. (2003). *What keeps teachers going?* New York: Teachers College Press.

Pillen, M., Beijaard, D., & Brok, P. D. (2012). Tensions in beginning teachers' professional identity development, accompanying feelings and coping strategies. *European Journal of Teacher Education*, 36:1–21.

Reeves, D. (2008). *Reframing teacher leadership to improve your school*. Alexandrina: Association for Supervision and Curriculum Development.

Shank, M. (2005). Common space, common time, common work. *Educational Leadership, 62*(8), 16–19.

Single, P., & Single, R. (2005). E-Mentoring for social equity: Review of research to inform program development. *Mentoring and Tutoring: Partnership in Learning, 13*(2), 301–320.

Smith, R. (2007). Developing professional identities and knowledge: Becoming primary teachers. *Teachers and Teaching: Theory and Practice, 13*(4), 377–397.

This chapter draws on the findings of a PhD study titled.

Tickle, L. (2000). *Teacher induction: The way ahead*. Buckingham: Open University Press.

Wenger, E. (1998). *Communities of practice: Learning, meaning and identity*. Cambridge: Cambridge University Press.

Yin, R. (2003). *Case study research: Design and method* (3rd ed.). Thousand Oaks: Sage.

Zuga, K. (1991). Technology teacher education curriculum courses. *Journal of Technology Education, 2*(2), 60–72.

Chapter 11
Technology Education Teachers' Professional Development Through Action Research

Tomé Awshar Mapotse

Abstract This chapter's purpose is to share the findings after engaging Technology Education (TE) teachers through an action research (AR) study. The chapter examines the claim that AR could be used as a vehicle for professional development of underqualified and unqualified TE teachers. The findings of this study indicate that AR impacted on TE teachers' classroom practice.

The Questions I Asked and Why I Think They Are Important

The subject Technology is referred to using diverse names in different countries. Some call it 'Science and Technology' (Malawi/Bangladesh), others label it as 'Design and Technology' (UK/Botswana/Namibia), yet other countries refer to it as 'Technology Education' (US/NZ/Australia), and in South Africa (SA) the subject is either called 'Technology' or 'Technology Education' (Mapotse 2015). There are other countries who gave the subject a totally different name; for instance, in Nigeria they call it Technical Education, and in Kenya it is pronounced as Art and Technology, whereas in Tanzania it is named Computer Integrated Education. I still believe that in other countries in the world it will be identified differently as well. The challenge experienced by most Technology teachers in SA is that they have been coerced into teaching this subject without any qualification (Mapotse 2012).

The situation for Technology teaching was exacerbated by a series of educational changes that have taken place in SA. These changes included the overhauling of the curriculum, followed by its review, which was a strategic and symbolic change since the first democratic elections in 1994. It was against this background that the main research question of the study was structured as follows:

> How could action research intervention be used to improve the teaching of senior phase Technology teachers who are unqualified and/or under-qualified?

T.A. Mapotse (✉)
University of South Africa (UNISA), Pretoria, South Africa
e-mail: mapotta@unisa.ac.za

P.J. Williams, D. Barlex (eds.), *Contemporary Research in Technology Education*, Contemporary Issues in Technology Education, DOI 10.1007/978-981-10-2819-9_11

The aim of this chapter is to provide readers with a perspective on the challenges facing Technology Education (TE) in developing countries today, as well as some possible skills and intervention strategies necessary to overcome these challenges, hence Technology Education teachers' professional development through action research. The extent of South Africa's unqualified and underqualified teachers in Technology has reinforced the notion that action research (AR) could be regarded as a tool for emancipation in the teaching of Technology.

Three methods were used during the pilot study for data collection, namely, observations, interviews and questionnaires. After data were collected from the three sources, I triangulated the findings. As Mokhele (2011: 99) explains, triangulation entails using multiple methods to find a valid, reliable and diverse construction of reality. It was during this pilot study that TE teachers raised the challenges encountered during their teaching of Technology.

The pilot study was the first phase (Phase 1) of my research for fact-finding purposes (Kemmis and McTaggart 1988). I then clustered the challenges from Phase 1 (see Fig. 11.1 in the section below) into themes to be responded to in the main action research roll-out. The main research question of the study sought to develop ways and means to enhance the teaching of Technology by teachers who are unqualified and/or underqualified to teach this subject.

In accordance with the main research question, the research sub-questions sought to find out more about the participants' Technology teaching:

- What is the nature of teaching Technology?
- How do senior phase Technology teachers at selected schools of Limpopo Province teach Technology?
- What are the AR-based intervention strategies that can be employed to improve teachers' Technology teaching?

The sub-questions above were outlined for the purpose of addressing identified challenges, which were developed into themes. These challenges suggested action research (AR) as an intervention to emancipate Technology teachers and improve their Technology teaching practice within the South African context.

How I Tried to Answer the Questions

In order to respond to both the main research question and the sub-questions, I divided my study into two phases as indicated in Fig. 11.1. Phase 1 was a minor phase which served as a catalyst to shape the study and consists of Cycle 1. This phase laid a solid foundation for further inquiry to unfold in the main AR Phase 2.

Phase 2 attempted to respond to the concerns and challenges raised by Technology teachers during the empirical reconnaissance investigation, and this is a major phase of the AR study. Phase 2 was unpacked through four spiral cycles (Cycle 2–Cycle 5). This phase was implemented later as the main methodology of the study, in which I met with the participants on a regular basis for AR cycles.

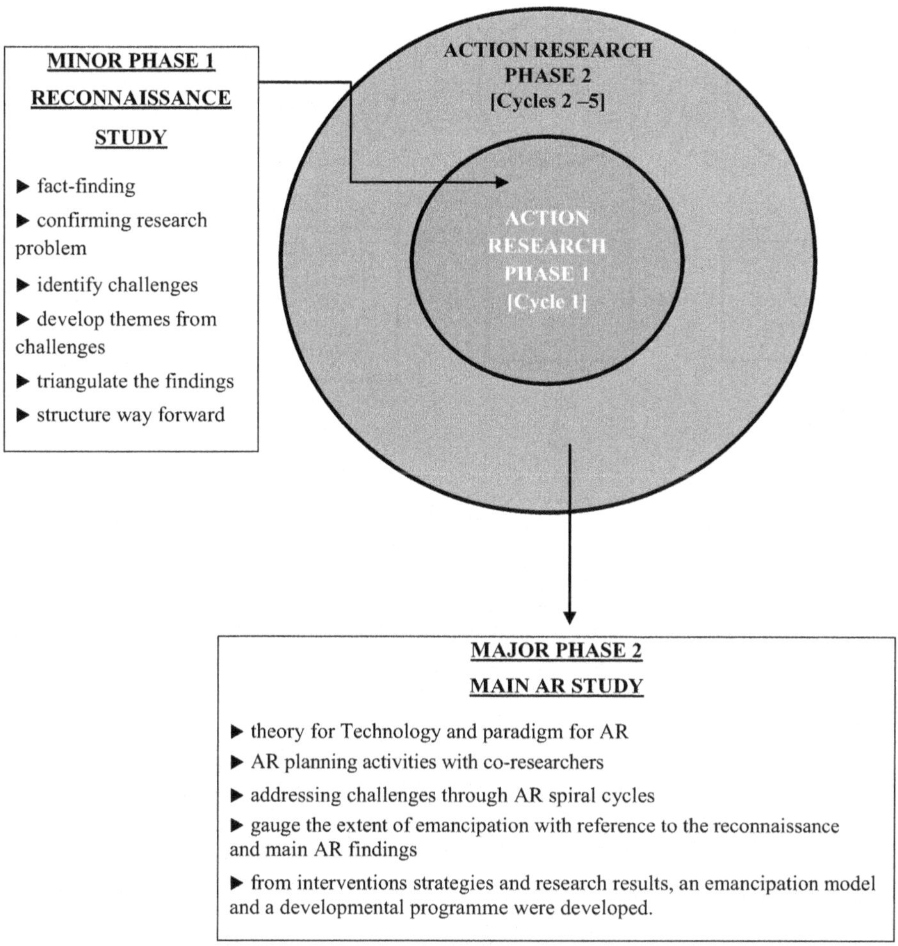

Fig. 11.1 Minor and major phases of the AR study

The contact or intervention session programme with the coresearchers was the key action to address the research questions. These intervention sessions with the participants were held every second month and were guided by the provincial work schedule of these TE teachers. The Technology themes as prescribed by the national ministry of education were followed each term based on the participants needs. Figure 11.2 illustrates Cycle 2–Cycle 5 of Phase 2 in addressing the research questions of the study and the needs or challenges raised by these TE teachers.

Figure 11.2 gives a summary of the AR contact sessions during Phase 2 with the coresearchers. All sessions were captured on both digital still and digital video for coding and analysis purposes. I formally invited participants to the sessions as a way to motivate them to attend. It was during Phase 2 that both the main research

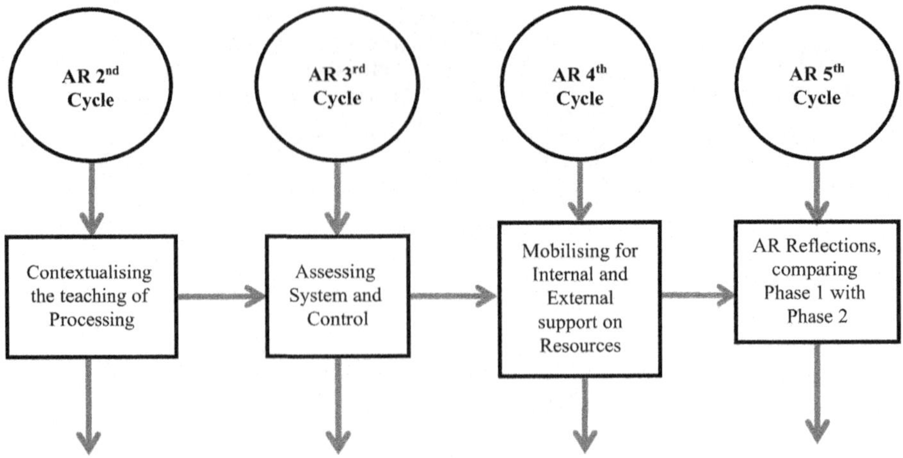

Fig. 11.2 Phase 2 intervention sessions of AR cycles

question and sub-questions were addressed with Technology Education teachers as AR coresearchers in Limpopo Province of SA.

Data were analysed by using the data sets approach. Johnson and Christensen (2004: 434) refer to 'data set' as a set of data the researcher uses in an attempt to convey the essential characteristics of data by arranging data into a more interpretable form. Mills (2000: 98) suggests that the next step in the AR process, after collecting data, is to reflect on what one has learned and summarise the meaning of data. I adopted the 'interim analysis' method during the implementation of my data sets. Interim analysis, according to Miles and Huberman (in Burke and Larry 2004), is a cyclical or recursive process of collecting data, analysing the data, collecting additional data, analysing those data, collecting additional data, analysing that data and so on throughout the research project.

What I Found Out

Findings from both Phase 1 and Phase 2 based on AR themes of the study and data set are highlighted in this section.

Findings of Phase 1

Findings from Observations

I photographically recorded all the classes being used to teach Technology in the selected five schools and found out that the teachers were not having any classes earmarked to teach only Technology. Teachers' classes that they were using for

Table 11.1 Biographical information of participants and their teaching experience

Gender		Technology teaching experience		Technology qualification		Context			Can plan Technology lesson	
M	F	Less than 6 years	More than 5 years	Yes	No	Rural	Semiurban	Urban	Yes	No
9	9	11	7	7	11	6	7	5	10	8

Technology tuition were also used by their colleagues to teach all other subjects. None of the classrooms had any Technology posters or displays of learners' projects. None of the schools had Technology workshops or laboratories in which to conduct Technology practicals, and no Technology kits or simple basic tools were available.

Findings from Interviews and Questionnaire

There were eighteen (18) participants in total from the five participating secondary schools, nine males and nine females. Eleven participants had less than 6 years of Technology teaching experience, while seven had more than 5 years. Eleven of the 18 had no form of Technology qualification and seven had some. Six worked in rural areas, seven taught at schools which are based in semiurban areas and five worked in urban schools. Ten could effectively plan Technology lessons, whereas eight were in need of some help. This account is displayed in Table 11.1.

In sharing the findings, attention will be drawn to the AR journey travelled with the coresearchers from Phase 1 to Phase 2.

Findings of Phase 2

Juxtaposing Action Research Phase 1 and Phase 2

Findings from both themes and data sets were juxtaposed in accordance with the 'before' and 'after' AR cycles. This is done to compare the Technology teachers' degree of emancipation throughout the cycles.

Figure 11.3 displays the AR journey travelled from Phase 1 to Phase 2 by the participants. The intention of displaying Fig. 11.3 is to map out the process of Technology teachers' emancipation from their early challenges to the stage where they felt confident with their knowledge of Technology and its pedagogy.

During Phase 1 challenges were identified and during Phase 2 they were addressed. From these challenges, themes were developed and strategies decided upon to address these themes as outlined in Phase 2 of Fig. 11.3. This is in line with what the Centre for Technology Education – Action Research (Mapotse 2012: 19) asserts, namely, that before one begins with intervention, one needs to gather baseline data.

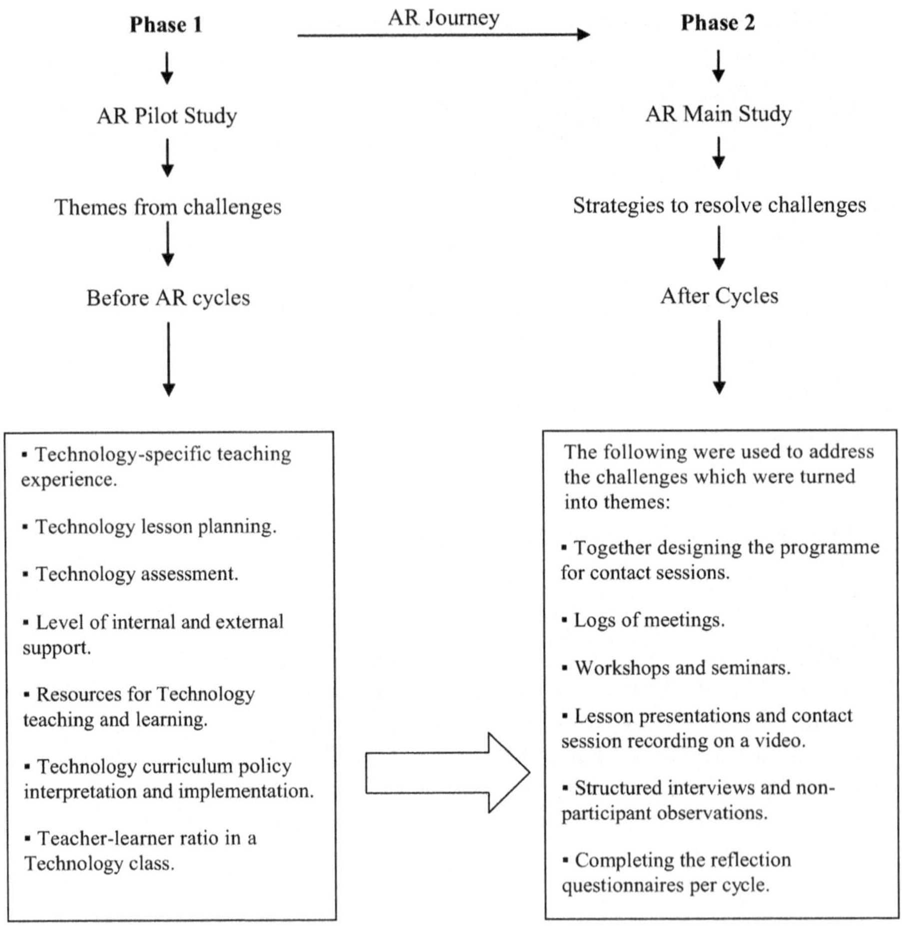

Fig. 11.3 Comparing findings between Phase 1 and Phase 2

It further stresses that knowing how participants performed before the beginning of the study provides a starting point for comparing the results.

Three data sources from the reconnaissance study indicated some challenges which confront TE teachers in their classroom practice. These challenges identified by the participants were reconfigured and clustered into themes as displayed in Fig. 11.3.

The findings and comparisons between preliminary and main study of the two phases are discussed in the next section.

Data Integrated from Three Data Sources (Observations, Interviews and Questionnaire)

The themes were selected to cover aspects of Technology teaching, from policy interpretation to classroom practice. These themes include Technology-specific teaching experience, Technology lesson planning, Technology assessment, level of internal and external support for Technology teaching, resources for Technology teaching and learning, Technology curriculum policy interpretation and implementation and teacher-learner ratio in a Technology class.

Technology-Specific Teaching Experience

Many teachers were asked to volunteer to teach Technology, and as a result, many of those teaching Technology did not have any qualifications in Technology Education. Some may be experienced in other subjects because they were already qualified in those subjects when they were asked to cross over into Technology. Given this and the findings from the teachers' biographical information in Table 11.1, there are teachers teaching Technology for reasons ranging from being coerced into teaching it to having a passion for it. For instance, in the interviews two teachers stated as follows:

It was just allocated to me.

It's fun, interesting and compels one to be innovative.

It was revealed during interviews that most of the Technology teachers are generally uncomfortable with the pedagogy of Technology. Some did not have any interest in teaching Technology, as one contended:

It just came along while I am already teaching and I didn't develop any interest in the subject.

It can be deduced from the teachers' responses that they were not grounded in the Technology subject, suggesting that the teachers encountered challenges in planning the Technology lessons.

Technology Lesson Planning

In the teachers' questionnaire responses, only seven out of eighteen teachers indicated the importance of using both the textbook and the curriculum policy document for their lesson planning. However, it surfaced from the interviews that teachers prioritised the use of a textbook:

If teachers were provided with at least a textbook we will be able to prepare our learning programme.

I don't think the challenges I meet as stated would have happened if I had relevant and enough textbooks for learners.

We need enough textbooks and learner support material.

It would appear that teachers expected to be supplied with a plan that would enable them to simply get into class and teach, without having to start by developing lesson plans:

We want to be supplied with pacesetters, scheme of work and draft lesson plans.

Only two of the teachers could produce their lesson plans. The lesson presentations were not without assessment challenges.

Assessment in Technology

Assessment should ideally be integrated with planning so that teaching and learning activities are not devoid of it (Mapotse 2014). An interview question sought to establish the assessment methods and evaluation procedures that Technology teachers applied during their teaching.

Some of the teachers mentioned the following assessment strategies that are applied in their Technology classes: assignments, class work, homework, tests and examinations. In the questionnaire teachers were asked to indicate if they planned any Technology design projects or any tasks (capability, case study or resource tasks) for their learners, but none of them did. One teacher blamed lack of support by the school:

Technology at our school is not taken into consideration because learners are not doing any practical work.

There are two possible reasons for teachers not to be engaged with any practical work with their learners: an inability to contextualise their Technology teaching and/or lack of resources.

Resources for Technology Teaching and Learning

It was observed that there was a lack of textbooks which augment but are not the only TE resource for both teachers and learners. In some schools there was not a single textbook among the learners for any subject. Some teachers in the same schools were sharing a textbook. This state of affairs was confirmed by the findings of interviews as expressed by respondents:

We don't have enough resources.
I guess Technology is a hands-on subject and there are no resources available.
Learners should be encouraged to buy necessary resources if needed.

The teachers pleaded with the Department of Education in Limpopo Province and their schools to provide the necessary resources for learners' group projects, a workshop centre where learners could do 'hands-on' technology, technology materials for demonstration and orientation and a technology resource centre.

Teacher-Learner Ratio in a Technology Class

Unmanageable learner numbers in a Technology class render the teaching of Technology ineffective (Ankiewicz et al. 2003). I observed that teachers' movement in the classroom during their lessons, and their interaction with learners, was extremely limited due to overcrowding. I had difficulty finding a chair to sit on during some class visits at some schools. The teacher-learner ratio ranged from 1:60 to 1:90 which had a number of consequences: even though the schools had a monthly schedule for tests, the interview findings indicated that the turnaround time for marking learners' written tests took between 2 and 3 weeks and so marking overlapped with the next test period. One teacher expressed his concern in this regard:

> *The department needs to improve the teacher-learner ratio so that the teacher is faced with a manageable class.*

The findings help the AR practitioner to structure the way forward with the coresearchers – TE teachers. We agreed on the frequency of our meetings for collectively addressing the challenges raised by the Technology teachers.

How This Might Be Used to Improve Teaching and Learning

Technology Education (TE) is unfamiliar to many teachers and a relatively new subject in school curricula both nationally and internationally (Mapotse 2015). The challenges that teachers face regarding the policy of Technology were articulated in terms of interpretation, analysis and implementation (Mapotse 2012). The situation for Technology teaching was exacerbated by many educational changes that had taken place in South Africa over the last few decades. These changes included the overhauling of curricula followed by a strategic and symbolic review which was also a sign of change since the first democratic election of 1994. These changes in the curriculum, revised twice, affected TE and demanded from TE teachers both knowledge of the subject content and pedagogical knowledge. The Ministry of Education in South Africa confirms by reiterating that while teachers in South African schools are qualified to teach a variety of subjects, many of the teachers of Technology are uncomfortable with the pedagogy of Technology (Department of Education 2003).

Challenges and Remedies for Teaching Technology

Teachers are generally faced with challenges in their teaching of Technology (Pudi 2007). In South Africa these challenges can also be understood with reference to the political transformation of the apartheid education system which has shaped the training and development of teachers in general. It is against this backdrop that

Gittings (1988: 6,16) postulates that no educational discussion in SA can be entered into without taking cognisance of the country's historical and political background. In the case of teachers moving out of an education system which was oppressive, it was imperative to intervene with AR, by incorporating critical theory as a way forward to collaborate professionally with, and emancipate, these TE teachers. Critical theory has an emancipatory intent (Mapotse 2012) and engages in real-world challenges. Both AR and critical theory may provide a structure for the emancipation of teachers from their situation as they are unqualified in the area of Technology Education.

Arising from this research and during my involvement with the AR participants in Limpopo Province, the following recommendations were proposed.

Advice to Technology Education Teachers

Based on my research and in the light of the work of others, I would advise Technology Education (TE) teachers to prioritise these three points to improve the quality of teaching TE:

- Let the TE teachers identify the university nearest their school and establish a relationship with Technology lecturers within the Mathematics, Science and Technology Education (MSTE) department so that they may request the university Technology lecturers to assist them addressing their core theme challenges.
- TE teachers should request materials from scrapyards, furniture repair stores and video and television repair stores as aids to consolidate their lessons and conduct demonstrations for their learners.
- TE teachers should organise technology-related excursions with their learners so as to expose their learners to the real world of Technology.

This may help the TE teachers to balance theory and practice and improve the quality of their lesson delivery. What Tooley (2000: 95) stated over a decade ago still holds today: 'In a fundamental dialectical relationship, theory and practice are indivisible'. This declaration makes sense since TE is a hands-on subject. TE core content areas and themes are based on a definition, which stresses that the outcome should be a product, artefact, model, ornament, new systems and/or processes. Pihama (1993: 40) once said: 'Cut off from practice, theory becomes simple verbalism. Separated from theory, practice is nothing but blind activism'.

Advice to District Officials

The following advice was formulated based on my research:

- The district should identify Technology teachers within a cluster who are knowledgeable about Technology content on certain core themes and let them be given the opportunity to empower their colleagues on cluster level.

- A circuit meeting should be arranged and questionnaires issued for Technology teachers to complete so as to identify gaps in their understanding of the TE curriculum.
- The questionnaires should be analysed and interpreted together with the cluster leaders.

Advice to Higher Education Institutions

The following advice was formulated based on my research:

- There should be a constituted forum of lecturers, national Technology facilitators (NTF), policymakers and curriculum planners that meet at least twice a year to reflect on policy and practice, identify gaps and come up with a plan to address them.
- Policymakers and curriculum planners should keep lecturers informed about what type of Technology teacher they should produce.
- The NTF should report on the provincial development of Technology, and the lecturers should report on their curriculum-aligned Technology modules.

Advice to the Ministry of Education

The following advice was formulated based on my research:

- Employ efficient and effective national Technology facilitators (NTF).
- Let the NTF make sure that each province has its own provincial Technology facilitator (PTF).
- Let the PTF surround himself/herself with regional Technology facilitators (RTF) with a TE background.

Assessment for Learning Technology

Assessment provides the feedback needed by the Technology teacher to successfully guide students' learning activities. If assessment is not manageable, feedback to students and other stakeholders will be delayed or the quality will be compromised. The teacher versus learners' ratio makes effective feedback to be a little bit difficult. De Dakar (n.d.: 29) summarises teacher-learner ratio as follows:

> The number of learners per teacher is an essential factor to be taken into account when defining the need for teachers. The Education for All Fast Track Initiative (EFA-FTI) framework recommends value of 40 learners per teacher in primary school.

It was evident from teachers who participated in this study that many confined themselves to assignments, class work, homework, tests and examinations as the most popular forms of assessment, which seemed to confirm my observation that teachers did not do any Technology projects or practical tasks with their learners. After AR was rolled out and after the need to contextualise the teaching of Technology Education was emphasised, the following additional assessment tools/methods were revealed and implanted here and there:

- Collage: can be done by learners if their teacher gave them a research task to investigate indigenous knowledge system (IKS)-related development, such as transport or communication.
- Projects: after teachers were conscientised to contextualise their TE teaching, they developed TE projects with their learners. The Grade 8 teacher introduced the indoor rubbish bin project and the Grade 9 teacher came up with test tube containers. All these projects were under the TE theme of 'containerisation'.
- Portfolio: each learner communicated all the steps he or she took in the form of a portfolio when they were constructing the grade-related project.
- Artefacts: the learners used waste material to construct their projects. Their artefacts were more real and ready to use.
- Rubric: teachers came up with a rubric to assess each project per grade.

The participatory emancipation for action research (PEAR) model is the summary of the AR activities that were addressing challenges by TE teachers of sampled schools raised during interaction with them from Cycle 1 to Cycle 5 of the study. The model summarises the journey travelled from Phase 1 (reconnaissance) to Phase 2 (main study) as outlined in Fig. 11.3 to emancipate TE teachers. Details of the PEAR model relating to data analysis are unpacked in the section that follows.

Knowledge Contribution of the Study

During planning, observations, actions and reflections of joint activities, the PEAR model to emancipate TE teachers, recommendations to improve teaching and learning of Technology, guidelines to develop teachers through action learning and a 6 week programme to develop TE teachers to enhance engagement with TE teachers were sought. Each step that we (the researcher and coresearchers) took was documented in the PEAR model. The positive results of all the steps and the journey travelled together were captured for recommendation purposes.

PEAR Model: A Way to Make a Difference in Teaching Technology

In response to my research findings, I have developed a model, naming it the PEAR model, which I believe will help individual Technology teachers collaborate and improve the teaching and learning in their classrooms. The model proposed for teacher emancipation is represented by the cyclical and spiral activities of action

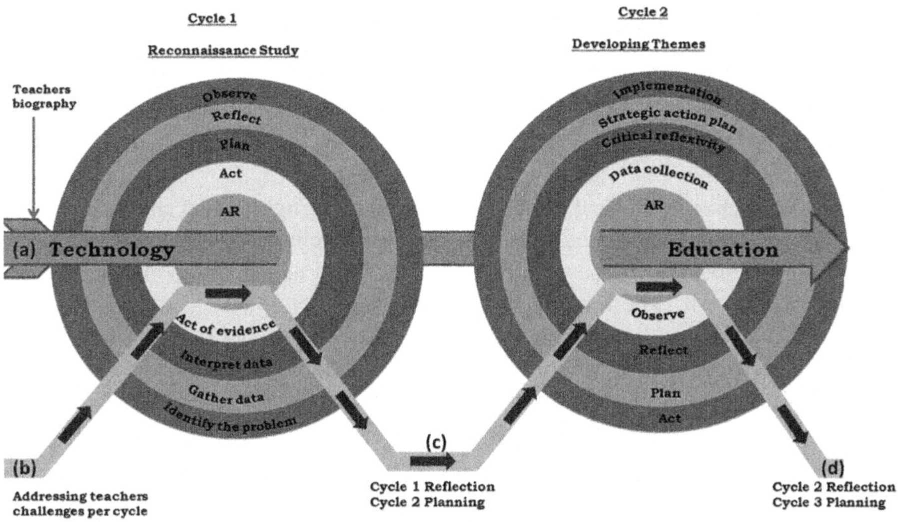

Fig. 11.4 The PEAR model

research, undertaken by Technology teachers in their classroom practices. My desire is that this participatory emancipation for action research (PEAR) model can be applied by Technology teachers in their individual classrooms. In this case the model will serve a dual purpose: (i) to help the teachers to assess their teaching and (ii) to facilitate learning. Teachers are advised to apply the model in the following manner (Fig. 11.4).

In Cycle 1: Teachers Should Embark on Target Population Discrimination of Their Learners Per Grade

- Before introducing a TE theme, let the teacher first assess what the leaner knows about the topic.
- The teacher is free to use any instrument for assessment purposes.
- During 'Cycle 1: Reconnaissance Study', the teacher should first observe learners doing TE activity as a group and identify the problem that the learners encountered.
- The teacher should reflect about the observations by gathering data from what was observed.
- The teacher should plan how to interpret the data gathered during observation.
- During the last stage, the teacher should act based on the evidence collected. The teacher then starts planning for Cycle 2.

In Cycle 2: Teachers' Continuation with the Model Application to Develop Themes

The AR spiral activities of observing, planning, acting and reflecting still repeat themselves in Cycle 2, but now there is a difference in the way in which the teacher as an AR practitioner interacts in leading the activities. Challenges (if any) identified by the teacher in Cycle 1 should be used as a springboard to shape his or her intervention. The teacher should list and code those challenges accordingly. From the coding, a choice of themes could be tabulated as a means of paving the way for further engagement. The teacher should strategically implement the action plan, observe the actions and, as the actions unfold, critically reflect on them. Lastly, the teacher should collect data and process it in the same way as in Cycle 1. Findings will feed into Cycle 3 to improve teacher performance and learners' achievement.

Recommendations to Improve Teaching and Learning of Technology

The research team went through the action learning (AL) spiral but on separate cyclical activities during this enquiry. Together we critically reflected on the outcomes per cycle from Cycle 1 to Cycle 5.

After data analysis from the cycles and interpretation of findings of the phases, I recommend the following regarding Technology teachers' professional development within a district:

- It all starts with one school to plan together for all Technology teachers per phase and follow both the work schedule and develop a learning programme collectively.
- Planning together will help teachers share the themes which are a bit challenging with one another and then seek expert advice.
- Once the teachers from a particular school emerge as a strong supporting team, they could invite a neighbouring school to form part of their TE group.
- Within a group formed by two schools, let the teachers identify the areas of expertise, and let those experts empower their colleagues within a TE group.
- Let the two schools agree on common TE assessment tools and projects for their learners; the group must develop a rubric per grade.
- Also encourage each other to practice peer assessment across the classes and/or schools.
- The group should be involved with AR evaluation strategies of observing, planning, acting and reflecting their process per term.
- The group should meet once at the beginning of the term to come up with a joint thorough plan of what they should commonly be teaching the next term.
- Once the two schools are doing well, they can invite another school to join them the following year; this means that every year the group adds a new school to

their team until a TE cluster is formed; this is what I call the Technology Education Cascading Theory driven by the expression, 'each-one-teach-one'.
- To culminate this process, the group should have TE competitions every 3rd term of the year where prizes are awarded to the best TE designed projects, products and systems including quiz responses for lower grades.
- All competitions should be contextualised and should solve the learners own community problems.

Guidelines to Develop Teachers Through Action Learning

Combining teacher collaboration with an action learning process is an effective way of providing support and guidance for action learners as they design their own action and engage in the process of studying the outcomes (Riel 2010). The guidelines below are earmarked for TE teachers to engage their learners in action learning (AL) before their lesson presentation:

- The teacher should start a knowledge-building dialogue with the learners based on the assumptions the teacher holds about the theme.
- The teacher should take into consideration that the learners are a community of diverse individuals from different backgrounds.
- The teacher should come up with mini projects within a bigger project and distribute leadership within the learners and hold the group responsible and accountable on deliverables.
- The teacher should carry out research work in learning circles – that is, a structure for organising group interaction within a set of guidelines to encourage individual ownership of the project.
- Both the ethical norms and expectations during the AL journey should be spelled out and highlight the learning circle product as it can be used as a stepping stone towards the final Technology product.

What Might Be Investigated Further

Six-Week Programme to Empower Teachers Through Action Research

The TE and AR programme that was followed to emancipate TE teachers emanates from the development of AR activities during the intervention stages with the participants. The programme can be executed in a minimum of 4 weeks, which means a week of contact session per term. Six weeks is the maximum period during which the facilitator could intervene and interact with the participants. Table 11.2 (below) highlights the 6-week intervention schedule for the action research practitioner together with those that need to be emancipated.

Table 11.2 Action research intervention strategies to empower Technology teachers

Week	Action	Cycle
One	Access, ethical observations and signing of consent	**1**
	Identify area of professional development or empowerment or emancipation. Embark on target population discrimination of your participants	
	Convince participants of the worth of action research	
	Conduct reconnaissance (include observation) study to confirm the research problem	
	Analyse data and prepare the findings	
Two	Share the findings, identify the challenges, convert these challenges into themes	**2**
	Plan together on how you are going to address these challenges	
	Be guided by the theory and the action research paradigm(s)	
	Prioritise those themes through action planning	
Three	Reflect on the action plan	**3**
	Identify those who can handle some challenges from the participants	
	Incorporate such as coresearchers and facilitate the process of addressing the challenges	
	Implement intervention strategies	
	Reflect on the cycles	
Four	Continue to implement action plan cyclically and spirally by observing, planning, acting and reflecting	**4**
	Reflect on the activities of the cycle	
Five	Let the emancipated participants display the sign of empowerment through learners' work	**5**
Six	Repeat what you have done during reconnaissance so as to confirm the degree of emancipation	**6**
	Analyse and crystallise data	
Seven	Do member checking and share the findings	**7**

Other Good Things to Share That Emanate from the Study

During member checking with the participants, the following consensuses were reached among AR coresearchers, their line managers (school management teams) and the district circuit manager that:

- The participating schools should have an annual budget for Technology.
- The TE teachers should not be shifted on yearly basis but be left to teach this subject for some years as this served as the basis for their growth, interest and development.
- The participants raised the issue that has been going on for sometime now that they didn't have any Technology subject advisor but the circuit manager responded by saying, during my 3 years of interventions and interaction, I served as their subject advisor.

After the study was concluded and I left Limpopo Province and went back to Gauteng Province where I am staying, the ensued incidents are also good to share:

- The circuit appointed the Technology subject advisor as it was recommended in my study.
- The newly appointed Technology subject advisor did draft a Technology empowerment workshop roster and used the emancipated participants of my study as co-presenters.
- The emancipated participants were elevated to be Technology cluster leaders.

In Gauteng Province where I am running a community engagement project with TE teachers of both Grades 8 and 9 from three secondary schools, I have implemented the guidelines through a 6-week programme to empower them as they were also unqualified and underqualified Technology teachers. Technology teachers need teaching models that can help them to approach Technology meaningfully and teach it effectively.

Technology teachers who participated in this study can now apply a variety of methods in their assessment. They managed to lobby for support from the district, their SMTs and parents. It is of great significance that they are ready to contextualise and do Technology projects with their learners by utilising available localised resources within the learners' community. The study developed and improved the Technology teaching practice of teachers.

How Might Teachers Contribute to These Investigations

Teachers taking part in activities similar to the one presented here will provide further useful evidence of the effectiveness of the PEAR model to enhance Technology Education.

Vignette of Both Activities and Findings from Cycle 3

The TE teachers in Grade 9 were for the first time encouraged to come up with a project for their learners. The teachers were keen to contextualise this project and motivate their learners to use waste materials for their projects, which resulted in an indoor dustbin project under the Technology theme, 'Processing', and the subtheme, 'Containerisation'. Photos 11.1 and 11.2 display a sample of the classroom dustbin project which was constructed by the Grade 9 learners of RMR Secondary School.

This product is made out of wire and a black refuse bag. The wire was used to construct the bin while the refuse bag material was used to wrap the wire. The bin can also be used as a teaching aid in addressing the theme: 'Systems and Control' » Mechanical Systems » Levers. The bin is opened by applying one's foot on the pedal so that internal mechanisms can open the lid. This bin further covers another

Photo 11.1 A closed rubbish bin

two aspects of TE, to which teachers can contribute, and that is the definition of TE in the South African context and its integration with Environmental Education. As the study is more focused on South Africa, the concept 'Technology' is presented from the DoE's perspective as follows: the use of knowledge, skills and resources to meet people's needs and wants by developing practical solutions to problems, taking social and environmental factors into consideration (DBE – Curriculum and Assessment Policy Statement 2011).

During member checking with the participants and their circuit manger, this rubbish bin constructed by learners was officially presented to the circuit manager as a gift. The presentation was a gesture of thanks to remind the manager about the AR journey once travelled within the Mankweng Circuit. This confirms that Technology teachers can teach the subject effectively and efficiently as long as they show some creativity and instil innovation in their learners. There is no need to lament about lack of resources, as recyclable materials can serve as a substitute while waiting for proper material from either the school or the provincial department of education. AR can be used for TE teachers' individual development.

An electronic copy of the PhD thesis can be found at this url: http://uir.unisa.ac.za/xmlui/handle/10500/7717?show=full

Acknowledgment My immeasurable gratitude goes to my Saviour Jesus Christ, who let wonderful people (too many to thank individually) to cross my path of knowledge acquisition. I want to express my indebtedness particularly to the following for their unreserved support:

- Prof M.T. Gumbo, my supervisor, mentor, friend and colleague.
- My dear wife Koena Catherine and my offspring: Phenyo, Thapelo, Kholofelo and Asanda
- Mr. Magagane, the Mankweng Circuit Manager and Limpopo Provincial Education Department.
- All participants and their SMTs in selected schools.

Photo 11.2 An opened bin from RMR Secondary School

References

Ankiewicz, P., De Swardt, E., Gross, E., & Reddy, V. (2003). The essential features of technology and technology education: A conceptual framework for the development of OBE-related programmes in technology education. *International Journal of Technology and Design Education, 13*, 27–45.

Burke, J., & Larry, C. (2004). *Educational research: Quantitative, qualitative, and mixed approaches* (2nd ed.). Boston: Pearson Education.

De Dakar, P. (n.d). *Universal primary education in Africa: The teacher challenge*. Breda: UNESCO.

Department of Basic Education. (2011). Education in South Africa [Online]. Available at: http://www.southafrica.info/about/education/education.htm. Accessed on 24 Mar 2011.

Department of Education. (2003). *Revised national curriculum statement grades R-9 (schools): Teacher's guide for developing learning programmes in technology*. Pretoria: Government Printers.

Gittings, J. (1988). Conference impressions. *Newsletter of Southern Association for Learning and Educational Disabilities, 7*(3).

Horkheimer, M. (1982). *Critical theory*. New York: Seabury Press.

Johnson, B., & Christensen, L. (2004). *Educational research: Quantitative, qualitative and mixed approaches* (2nd ed.). Boston: Pearson.

Kemmis, S., & McTaggart, R. (1988). *The action research planner*. Victoria: Deakin University Press.

Mapotse, T. A. (2012). *The teaching practice of senior phase technology education teachers in selected schools of Limpopo Province: An action research study*. Pretoria: DEd, University of South Africa.

Mapotse, T. A. (2014). An emancipation paradigm through critical theory in technology education: An action learning paradigm. *Mediterranean Journal of Social Science. 5*(3), 484–493. ISSN No: 2039–9340 (print)/ISSN No: 2039–2117 (online). Web: http://www.mcser.org

Mapotse, T. A. (2015). An emancipation framework for technology education teachers: An action research study. *International Journal of Technology and Design Education, 25*, 213–225. doi: 10.1007/s10798-014-9275-y, Springer Publishers. ISSN No: 0957–7572 (print)/ISSN No: 1573–1804 (online). Web: http://download.springer.com

Mills, G. E. (2000). *Action research: A guide for the teacher researcher*. Upper Saddle River: Prentice-Hall.

Mokhele, M. L. (2011). *Teachers' perspectives on continuing professional development: a case study of the Mpumalanga secondary science initiative (MSSI) project*. DEd thesis, University of South Africa, Pretoria.

Pihama, L. (1993). *Tungia te Ururua Kia Tupu o te Harakeke: A critical analysis of parents as first teachers*. MA thesis, University of Auckland.

Pudi, T. I. (Ed.). (2007). *Understanding technology education from a South African perspective* (pilot ed.). Pretoria: Van Schaik.

Riel, M. (2010). *Understanding action research*. Centre for Collaborative Action Research. Pepperdine University. Available at: http://cadres.pepperdine.edu/ccar/define.html. Accessed on 20 Oct 2014.

Tooley, W. R. (2000). *Political rationality & government mechanisms: Maori education policy in the new millennium*. MA in Education dissertation, University of Auckland, Auckland.

Chapter 12
Technology Education: Education for Enterprise (E4E) in New Zealand (A Connected Curriculum)

Gary O'Sullivan

Abstract The research reported in this chapter was underpinned by the interpretive-constructivist paradigm utilising a Fourth Generation Evaluation methodology. The purpose of the research was to examine conflicting rationales for the implementation of technology education and Education for Enterprise and to evaluate a professional development project.

Introduction

Two central components of the research discussed in this chapter are technology education and Education for Enterprise. The former is a mandated curriculum offering, and the latter is an approach to learning which encourages connectivity and gives students responsibility for their learning. Both of these components are reflective of a New Zealand culture and psyche which appreciates and acknowledges creativity and innovation.

The current New Zealand Curriculum (NZC) was published late in 2007. The process of technology in the curriculum is described as:

> *intervention by design: the use of practical and intellectual resources to develop products and systems (technological outcomes) that expand human possibilities by addressing needs and realising opportunities. Adaptation and innovation are at the heart of technological practice. Quality outcomes result from thinking and practices that are informed, critical and creative.* (Ministry of Education 2007, p. 32)

Education for Enterprise in New Zealand has been defined in broad terms by a number of key stakeholders as:

> *... a teaching and learning process directed towards developing in young people those skills, competencies, understandings, and attributes which equip them to be innovative, and to identify, create, initiate, and successfully manage personal, community, business, and*

G. O'Sullivan (✉)
Private Educational Consultant and Entrepreneur, Manawatu, New Zealand
e-mail: nzconsult4u@gmail.com

P.J. Williams, D. Barlex (eds.), *Contemporary Research in Technology Education*, Contemporary Issues in Technology Education, DOI 10.1007/978-981-10-2819-9_12

work opportunities, including working for themselves ... It is about how we teach across the curriculum and how we get our students to take ownership of their learning. Education for enterprise is not a discrete subject but provides learning experiences that encourage young people to be active participants in their learning. (Ministry of Education 2009, para.1)

Although developed separately, these component definitions display a number of similarities; these are explored in this chapter within the notion of a connected curriculum.

The Questions I Asked and Why I Think They Are Important

The purpose of the research reported in this chapter was to evaluate a technology education and Education for Enterprise professional development programme designed to bring together school and community partners. To facilitate this, five questions were developed:

Question 1: What recognisable ideologies underpin technology education, enterprise education and school-community links in New Zealand?

Young (1998), when talking about flexible specialisation and its relevance to education, introduced a notion of connective specialisation. This contrasted with the insularity of traditional discipline specialists and ultimately with the divided curriculum. Young argued for a shift from teacher centredness to learner centredness. This could be seen as an argument between discipline and interdiscipline. Muller (2011) outlines a discipline as a specialised, self-regulated domain of knowledge and practice. Alton-Lee's (2003) research suggests learning is most effective when 'curriculum enactment has coherence, interconnectedness and links are made to real life relevance' (p. viii). Technology education should be a broadly connected curriculum with a strong emphasis on developing practical creativity, enterprise and innovation. That is not to say that the discipline area be discounted; rather I advocate that students should be given the opportunity to develop both facets: discipline understanding (depth) and connectedness (breadth).

Question two was designed to explore this hypothesis, specifically through interactions between myself as researcher, the facilitators and the participant teachers in the particular research project.

Question 2: What is a creative connected curriculum? The thesis discussed in this chapter focussed on Education for Enterprise as an outcome. According to Gibb (2008), being enterprising is a conative and affective as well as a cognitive endeavour. This required all participants to revisit their conceptual understandings and make adjustments to their pedagogy to fulfil the educational intentions.

Question 3: What teacher practices support or undermine development of enterprising attributes within the Education for Enterprise project?

My teaching experience as well as other research projects I have been involved with had led me to the realisation that constructive undertakings which rely only on the endeavours of particular individuals fail to have any longevity and therefore any real significant educational impact.

Question 4: What school-wide practices support or undermine development of enterprising attributes within the Education for Enterprise project?

Throughout my educational career, I have been involved in community partnerships, and this has helped shape my understanding of the role technology education plays in education connectedness. The New Zealand Ministry of Education, through its policy directives (http://nzcurriculum.tki.org.nz/Principles/Community-engagement), outlines why these school-community partnerships are required.

Question five addressed the impact of school and community partnerships or links on the Education for Enterprise professional development programme.

Question 5: What are the influences and impacts of school-community partnerships on teaching and learning within the Education for Enterprise project?

How I Tried to Answer the Questions

Throughout the research project, my role had been more than just a researcher; there had been times where I had been a teacher, facilitator and learner. This has been an epistemological exercise in which all stakeholders, including both evaluators and participants, have had multiple roles. This I believe led to an established, sophisticated trust where claims, concerns and issues could be expressed and critiqued openly.

In this research, an interpretive-constructivist paradigm as described by Denzin and Lincoln (2000) was employed. Under the interpretive-constructivist paradigm, interactions between all stakeholders, including the researchers, are deemed equally important. These interactions, combined with an exploration of values held by all the stakeholders, help shape the information which became a major focus of the study. The interpretive-constructivist paradigm can be characterised by its use of primarily qualitative data-gathering techniques in a hermeneutical and dialectical manner. Interpretive-constructivist researchers focus on the multiplicity of viewpoints held and illuminate how these interact to shape the study (Denzin and Lincoln 2008). It is the mutual interactions between those studied and those doing the studying that guide the research outcome. The interpretive-constructivist paradigm supports the view that the observed reality exposed as part of a research study is the result of a social construction process and that no single truth is discoverable.

In the case of the specific research reported in this chapter, defining and making sense of the impact of a professional development programme needed to be co-constructed from the perspectives of all the participants. Interpretivist methodology is a participative and collaborative endeavour concerned with constructing new understandings. Fourth Generation Evaluation (Guba and Lincoln 1989) was used as a method to investigate the Education for Enterprise professional development programme for teachers. It was selected because this method focuses on negotiation (the hermeneutic/dialectic). The tensions and conflicts associated with this methodology are important when discussing the role of active participants and ownership of evaluation studies.

With the New Zealand Curriculum (2007) embedded and the Education for Enterprise professional development project concluding, it is timely and important for academic research in this area to be reported. Connecting school activity with out-of-school experience is not a new concept; it relies heavily on the work of Dewey (1933) and reflective thinking.

The study involved teachers from 16 schools clustered in three regions of New Zealand. Both primary and secondary school teachers participated in three rounds of four workshops spanning a 3-year period. All four workshops were repeated in each of the three geographical and population clusters – coastal, rural and city. Most of the workshops were attended by all three of the Education for Enterprise project team consisting of myself as a researcher and two experienced facilitators. The workshops were conducted as a partnership between the 'local' experts (the teachers) and 'non-local' experts (the facilitators and researcher). The focus was on 'mutual aid' provided to improve teaching and learning of 'Enterprise for Education' through the learning area of technology. Data were generated about the teachers' knowledge, thinking and practice through questionnaires, semi-structured interviews, classroom observations and reflective journals. Observation transcripts were analysed, coded, discussed and reflected on during reflection blocks at the beginning and end of each workshop.

This research was designed to investigate the effects of a professional development programme on the quality of Education for Enterprise and technology education interactions in classroom learning environments. It also provided the participating teachers with an extended period of reflective evaluation. Because it was based in New Zealand school classrooms and involved regular classroom teachers, the study has provided valuable, in-depth information about everyday classroom practice. Early staff meetings, syndicate meetings and workshops focused on the ministry's definition for Education for Enterprise and the recognition and promotion of the enterprising capabilities. Facilitators gathered baseline data in each school and wherever possible found links in documents, including school vision plans, planned pathways of learning and strategic plans that showed the school had already initiated aspects of an enterprising approach or wished to follow these directions.

The professional development project aimed to provide the participating schools with opportunities to develop professional learning communities, i.e. communities of practice (Wenger 1998). These would be comprised of school staff, the external facilitators who supported the school staff and a project researcher. Together, this community would develop and reflect on their professional knowledge and practice of Education for Enterprise within technology education. According to Holly and McLoughlin (1989), professional development can be viewed as a major factor in successful efforts to improve schools. The improvements here were to be identified by shifts in understanding and practice. These shifts would be encouraged by utilising an extended programme of facilitation action and reflection.

What I Found Out

According to Guba and Lincoln (1989), claims, concerns and issues form part of a conversational process to allow stakeholders to render judgements about the evaluation that is taking place. The Education for Enterprise team challenged the participants to review and possibly alter their practices where necessary to allow for closer synergies between technology education, Education for Enterprise and community involvement.

The findings of the study revealed that the main point of difference between Education for Enterprise and other quality learning is that successful Education for Enterprise is based on authentic local needs that students perceive from the start of the project. These perceived needs help to make the learning connected and often involve working with community partners. Activities that work well are those that are linked to practical undertakings and include tangible outcomes such as those found in technology education. Instances where the students failed to be supported by a teacher/facilitator invariably led to a failure of the experience and outcome.

The research indicated that a combination of technology education and Education for Enterprise was successful when students were given a controlling function within the project. This led to enhanced ownership of their individual learning. This controlling function is enhanced if time is given in the early stages of the unit to build students' knowledge. Many teachers and students struggled with these early stages, known in technology literature as the fuzzy front end, Rhea (2005).

By the end of the final workshop, nearly all teachers had appreciably improved the quality of their classroom offerings in relation to their inclusion of Education for Enterprise. The research demonstrates the effectiveness of quality supported development programmes and responsive evaluation research as a model for reflective professional development and enhanced teaching and learning.

Support at the Classroom Level

In summary, there has been a documented growth in teachers' professional understanding. There is evidence to support Education for Enterprise as part of technology education when:

- The context for the activity is shared, authentic and real.
- The activity is linked to practical undertakings and includes tangible outcomes.
- Students are given a controlling function within the project, which leads to ownership of individual learning.

- The student contribution is encouraged, mentored and acknowledged.
- Students are provided with flexible frameworks to facilitate project management.
- The approbation of Education for Enterprise is evident from the teacher.
- The teacher reflects on their delivery and involvement from the beginning to ensure an enterprising approach is taken and modelled.
- The teacher encourages and values reflection from the students, and this is incorporated in progression and assessment.
- Time management is identified as important; therefore, the teacher might operate both as a facilitator and a goal chaser.

Case studies from this project have been published on the MOE Education for Enterprise website http://education-for-enterprise.tki.org.nz/. Those collected and discussed at the workshops demonstrated a range of teaching strategies that teachers had put in place to enable their students to work in authentic contexts to perceive relevance and purpose in what they do and how they connect to the wider world. This chapter highlights planning notes and frameworks being used in classrooms and schools that evidenced the development of enterprising attributes in students. This research has demonstrated that an enterprising culture is developing in the schools involved in the project, and the number and range of strategies trialled and reported on through the observations and recorded interviews had been substantial.

Support at the School Level

As above, there is evidence to support Education for Enterprise as part of technology education when:

- Support for participation and monitoring comes from the senior management team of the school.
- Participation is shared to prevent burnout and remains consistent in the event of staff changes.
- The understandings of Education for Enterprise are shared amongst all staff, not just those involved in particular projects.
- Education for Enterprise is not seen as another extra but is interwoven with key learning intentions from numeracy, literacy and the technology curriculum area.
- Time allocations are flexible enough to allow for appropriate research and enough time to see the projects through.
- Time allocations are concentrated, not too disjointed or disrupted by other activities within the school (suspended timetable alternatives).
- Consideration is given on how to place Education for Enterprise within the school or whole school.
- Recognition and utilisation of quality facilitation and advisory programmes are given.

Concerning School-Community Partnerships

Of the 106 units of learning involving an Education for Enterprise approach in this research, the education facilitators assessed that:

- 57.5 % involved a high level of community partnership.
- 18.8 % involved a medium level of community partnership.
- 23.7 % involved a low level of community partnership.

The assessment was based on observations and discussions with the teachers involved. Based on these experiences, there is evidence to support Education for Enterprise as part of technology education when:

- Boards of trustees and parents are involved in the planning stages.
- Experts and mentors are sought as soon as possible, and their time commitment is established.
- Cooperation and coordination of assistance occur between mentors and teachers.
- Due consideration of student participation and decision-making is kept central to the project.
- There is community pride established in the activities undertaken.
- Reports and updates are provided to all parties regularly using a variety of media.

There is an indication that, where successful, a quality Education for Enterprise approach as part of a technology education programme can help improve:

- Behaviour management and motivation
- Participation of students, especially where their learning styles do not always suit the traditional expectations of a classroom
- Stronger ties between the school and the community, through a connected curriculum
- Meaning and therefore a better learning experience to aspects such as numeracy, literacy and developing specific curriculum knowledge

According to Stoll et al. (2006), there are five characteristics of professional learning communities, and each was important to this project:

Shared values and beliefs – such beliefs would impact on both decision-making and action.

Collective responsibility – this helps to sustain commitment to the project for the lifetime of the community.

Reflective professional enquiry – this requires observation and analysis joint planning and development.

Collaboration – this requires identification of shared purpose, activity and achievement.

Finally, a realisation that *group as well as individual learning is to be promoted.*

Wood (2007) also recognises the importance of collaboration through the active participation of teachers in the development and sharing of knowledge in professional learning communities. Wood describes a Deweyan approach utilising collective inquiry through systematic observation and analysis of classroom activity as the basis of professional learning. Technology teachers are going to have to be active in questioning the perception and understandings of all the community parties involved.

The research questions discussed in this chapter were developed to try and identify and then facilitate shifts in understanding and practices in the participating schools with regard to Education for Enterprise as part of technology education and meet the five characteristics identified by Stoll. To facilitate this, there was some recognition of Waters et al. (2003), assertions that any shift requires action at both individual and school level and would involve second-order change, i.e. shifts in practice that require an examination of personal beliefs and a new way of working.

Connecting curriculum integration via the technology curriculum with the wider community was also an important aspect of this study. Not all attempts to include wider involvement were successful. Some of the frustrations experienced during community participation in this research project included:

- Enterprise mentors taking too much control over the direction of work.
- Lack of response from community experts.
- Mentor's/expert's timeline and school timeline not matching.
- Expert's work commitments during school time obviously taking precedence.
- Experts wanting to relax from their own job in their own time. There was willingness but commitment issues came to the fore.

The findings support notions of authentic situations, shared ownership and the integration of Education for Enterprise within existing curriculum areas such as technology education. Teachers, with support from senior managers and facilitators or advisors, can make useful connections with the wider community to enhance their teaching, and ultimately the students' learning that occurs has more meaning. According to Hill (1998), involvement with business and the community through technology education is essential to provide authentic educational experiences for our young people if they are to cope in the new millennium. Closing the gap between school life and workplace life is an important step in this direction.

Consistent with claims made by Mitchell and Cubey (2003), the analysis of data in the present study was facilitated by my involvement. This active participation is also considered a common facet of Fourth Generation Responsive Evaluation.

How this Might Be Used to Improve Teaching and Learning

Atkins (as cited by Dykes 2002) claims that a focus on beliefs and challenging these is a critical component of effective professional development if it is to lead to lasting and significant change. School-based professional development that includes this focus relates to the classroom; it is designed to be appropriate to the teacher involved; it is student orientated and involves a whole-school approach that can ultimately lead to successful development (Pratt, Lai, & Munro as cited by Lai 2001). Effective professional development also needs to be ongoing, and thus the professional development in this project was planned to spark discussions about effective planning and pedagogies. These discussions are identified as the first step towards having a group vision that could be owned and supported by all teachers (Alexander et al. as cited by Lai 2001). The second step was the provision of tools to put this vision in place.

As a result of the professional development programme, the teachers' planning incorporated a new focus of developing authentic contexts and the meeting of a real-world context. Planning templates and process guides that assist this planning were used and adapted. With over 100 examples of enterprising concepts being woven into classroom programmes of work, there was evidence that suggests that teachers had a growing understanding of the role and value of Education for Enterprise attributes and capabilities. Responses from a data-generating strategy in which participants were asked to edit the MOE definition of Education for Enterprise suggested that they had developed a more positive understanding of the value of Education for Enterprise and that the learning environment should stimulate activity and autonomy.

This research indicated that a combination of technology education and Education for Enterprise was successful when students were given a controlling function within the project. This led to enhanced ownership of their individual learning. To facilitate this, the teachers' role changes from one of knowledge provider to one of helper or guide. The teacher facilitates control of the project by the students and allows them to make their own path to the answer, rather than the teacher laying the path out for them. Gentry (2000, p. 11) states 'Teachers are the channel through which the students acquire the skills for learning…not as a supplier for knowledge but as a prod for students to gain that knowledge on his or her own'. According to Torp and Sage (2002), students respond to obtaining this control by having a personal investment in the solution. Teachers can develop the controlling function in their students by teaching strategies which help students to:

- Identify and clarify appropriate contexts and activities
- Develop their communication skills
- Improve time, resource and team management capabilities
- Develop their action planning and evaluation abilities

One example of this discussed in the research was a project which highlighted the relationship between a school and the local fire brigade which was having a 50-year Jubilee. The brigade was restoring their original 1932 V8 fire engine, and students followed the progress and created a website to highlight the restoration and promote the jubilee. The students also had input into the design of the museum being built at the fire brigade. The students were introduced to the project with a trip to the restoration site, and they were able to interview and work closely with the restoration team. Another example reported was a school working on a unit called weedbusters which was adapted from a Department of Conservation unit on noxious weeds. New Zealand has about 2500 species of native plants, but ten times as many introduced exotic plants, the school involved had a recently cleared area and could see how weeds easily took over if left uncontrolled. They worked with a local garden centre evaluating existing tools for weed extraction and worked in teams to design, model and promote alternatives to those currently available. Considerations such as ergonomics and hand protection came to the fore.

One of the concerns reported from phases one and two of the study concerned the 'lack of time' teachers and students had for the action/doing part of Education for Enterprise and technology education. This concern was the basis for an emergent research question: how do we find the time for an Education for Enterprise activity within technology education? In response to the concern raised about a lack of time, I devised a professional development session which identifies how this issue is addressed by the wider technological community. The fuzzy front end is how practicing technologists describe the confusing early establishment stages of a design or problem-solving process. Typically this is the stage where the technologist is trying to establish what the real need, want or opportunity is, followed by asking how they might begin to address or solve the problem. Often it involves a particular type of cognition and an iterative process that can lead to early ideas being discarded for a number of reasons. These can include viability, cost-effectiveness, originality and functionality.

I showed the importance of the fuzzy front end and the planning and preparation stage by relating actual examples from industry. Teachers found that taking time in this area was an important part of the technology process and that they shouldn't worry if it takes them longer than they think it should. In the real world, this fuzzy front end can account for 50 % of development time. The teachers developed the understanding that successful Education for Enterprise is achieved via practical activity, that is to say that enterprising initiatives are encouraged through engaged activity such as modelling and outcomes which are tangible in nature. This understanding helped teachers to make connections between Education for Enterprise and technology education. It was also noted that students should be given more responsibility within the projects with the teacher acting as an encouraging advisor. This could be seen as supportive of a Deweyan philosophy where the learning is a complex endeavour that can be characterised by a connective relationship between knowledge, action and the authenticity and value of the activity. Accordingly, there should be an atmosphere of empathy, trustworthiness and encouragement between those involved. Many experts in the field of evaluative professional development (Hammerness et al. 2005) agree that unless teachers are assisted to develop their reflective skills to the point where they are able to critique and monitor their own behaviour in the classroom, routinised and unreflective practice will be unlikely to change. This was exemplified during this project with many of the teachers initially unable to allow students to manage their own design process. It was felt that projects would fail if the students were in control. This view changed for many due to the professional development workshops.

What Might Be Investigated Further

According to Guba and Lincoln (1989), claims, concerns and issues form part of a conversational process to allow stakeholders to render judgements about the evaluation that is taking place. The Education for Enterprise team challenged the participants to review and possibly alter their practices where necessary to allow for closer synergies between technology education, Education for Enterprise and community involvement.

I felt confident that participants had grown sufficiently comfortable with each other and us to be open and honest in their responses. The importance of this trust issue is highlighted by Krueger and Casey (2000) who contend that rich data can only be generated if group individuals are prepared to engage fully. Kitzinger (1994) suggests that feedback obtained from participants who see each other as acquaintances are more willing to challenge each other's ideas. My role as researcher was to try and make sense of the individual responses, and also to be analytical enough to see any developing relationships between the responses, and the research objectives as a whole.

The development of successful school-community partnerships identified in this research suggests that they are dependent on numerous factors. First, the community involvement must be sought at the planning stages of the activity. Second, partnership members need to be identified. These could include boards of trustees, parents as well as experts and mentors from the community. Third, it was identified as important to establish time expectations and commitments as early as possible. Fourth, teachers are likely to need to develop key roles within the partnership. Establishing learning intentions and ensuring student participation and their control over decision-making should be kept central to the project. Careful planning is required to ensure that timely involvement is achieved and that the teacher must maintain community-partner focus to ensure learning intentions are met. Fifth, they might try to establish a collective community pride in the activities undertaken – this can be maintained when reports and updates are provided to all parties regularly using a variety of media.

There has been an emphasis on authentic needs while designing the units and on making these needs transparent to the students so they identify where their new found skills and knowledge will be utilised. There was also an expectation for students to create more than just presentations at the using and doing stage by trying to push on and develop tangible outcomes. The Education for Enterprise attributes had been a central feature when designing the whole unit rather than simply being treated as an afterthought. Curriculum integration can create more time and reflect the integration that occurs in real life. The project team was able to ensure that the students reflected on and revisited the timeframe, deadlines and expectations continuously. Success in Education for Enterprise requires teacher scaffolding that will help students move beyond the fuzzy front end. Teachers must help to develop the background knowledge required by getting students to ask relevant questions. As well as this, teachers should maintain a focus and continuity in terms of timetabling and commitment. Student involvement in planning, goal setting and identifying tangents where student ideas are actually considered and used has also emerged as a theme from the research.

The professional facilitation had at its core a problem-based learning instructional strategy (Ward and Lee 2002). This had been identified by the Education for Enterprise project team as a way to allow the students the opportunity to (a) become authentic stakeholders in their learning; (b) identify key facets of science, technology and curriculum integration; and (c) participate in a learning environment which modelled a useful classroom approach.

By the end of the project period, there had been a shift in the nature of activities undertaken by the participating teachers. Their growing understanding of both Education for Enterprise and technology education was reflected in the outcomes developed from the units undertaken. Although the professional development project demonstrated that bringing about changes in teachers' practices can be a slow and complex process, the systematic experiences provided within the four phases of this evaluative research eventually proved to be effective in assisting the teachers to develop the requisite knowledge and reflective skills to bring about change in their practice. Involving the teachers in the process of generating data from within their own classroom settings was a powerful catalyst that facilitated a robust evaluation process and ultimately changes in their programme offerings.

With this type of project evaluation, there is always the question of bang for buck. This economic imperative has been somewhat alleviated by publication of case studies from this project on MOE-funded websites. For some examples, visit http://nzcurriculum.tki.org.nz/Curriculum-resources/Education-for-Enterprise/Resources.

How Teachers Might Contribute to These Investigations

The limitations of evaluating any individual Education for Enterprise programme are highlighted by Lewis (2002), who asserts that 'all enterprise education programmes have different characteristics and are delivered in a variety of different contexts (cultural and educational)' (p. 21). This research was undertaken in particular regions of New Zealand, and there will always be questions of national and indeed international transferability. That said effective teacher engagement in and with research has helped us to understand that the ways teachers learn may be more like the ways students learn than we have previously understood. Learning theorists such as Resnick (1989) and Schon (1991) have indicated that people learn best through active involvement and by thinking about and articulating what they do.

A teacher, in cooperation with the senior management team of the school and in consultation with any health and safety officer from the community partner, must maintain a safe operating environment. It is important to ensure that both schools and the partnership have shared understandings of, and agreed targets for, the learning outcomes. To maintain integrity and continuity, teachers should link pre-partnership learning activities with any site visits and follow these up with post-visit activities. Discussions could occur about the partnership with students to ensure they appreciate that the enterprise people are giving up their expensive time. Students should try to be punctual, polite and prepared. Teachers can facilitate this by matching the partnership to the age and learning level of the students involved. Additionally they could appreciate that they are representing themselves, the school and the integrity of the partnership for future participants.

One of the phrases we used with the participants was *being brave.* We have all had opportunities in our teaching careers to try something different. Given the opportunity, I would advocate being brave and encourage community partnerships. This study proved to be successful in enhancing teacher knowledge and understanding of Education for Enterprise and technology education. Participation in research around teacher practice and evaluative studies of professional development is essential if we are to establish ways of making improvements. If you are active in trialling new approaches, please make contact with educational researchers who can help with formalising data gathering and dissemination of the findings. The MOE in New Zealand has helped to establish a number of regional clusters for E4E, and their role is to work with schools, businesses and community groups to enhance E4E understanding and develop E4E partnerships between schools and communities/businesses. Other organisations such as Futureintech and the Young Enterprise Trust also offer support materials for teachers looking to establish E4E activities.

If you are interested in pursuing this time of activity for your students, visit their websites; they are full of useful material to support your endeavours http://www.futureintech.org.nz/ and http://yetrust.co.nz/.

References

Alton-Lee, A. (2003). *Quality teaching for diverse students in schooling: Best evidence synthesis.* Wellington: Ministry of Education. From http://www.minedu.govt.nz/web/downloadable/dl11124_v1/quality-teaching-for-diverse-students-in-schooling.doc

Denzin, N. K., & Lincoln, Y. S. (Eds.). (2000). *The handbook of qualitative research* (2nd ed.). Thousand Oaks: Sage.

Denzin, N. K., & Lincoln, Y. S. (Eds.). (2008). *Strategies of qualitative inquiry* (Vol. 2). Los Angeles: Sage.

Dewey, J. (1933). *How we think. A restatement of the relation of reflective thinking to the educative process* (Revised edn.). Boston: D. C. Heath.

Dykes, R. (2002). Information literacy – A separate skill or a new approach to learning? *Computers in NZ Schools, 14*(2), 26–30.

Gentry, E. (2000). *Creating student-centered, problem-based classrooms.* Huntsville: University of Alabama in Huntsville. Retrieved March 20, 2009, from http://aspire.cs.uah.edu/

Gibb, A. A. (2008). Entrepreneurship and enterprise education in schools and colleges: Insights from UK practice. *International Journal of Entrepreneurship Education, 6*(2), 48.

Guba, E. G., & Lincoln, Y. S. (1989). *Fourth generation evaluation.* Thousand Oaks: Sage.

Hill, A. M. (1998). Problem solving in real-life contexts: Alternatives for design in technology education. *International Journal of Technology and Design Education, 8*(3), 203–220.
Hammerness, K., Darling-Hammond, L., Bransford, J., Berliner, D., Cochran-Smith, M., McDonald, M., et al. (2005). How teachers learn and develop. In L. Darling-Hammond & J. Bransford (Eds.), *Preparing teachers for a changing world: What teachers should learn and be able to do* (pp. 358–389). San Francisco: Jossey-Bass.
Holly, M. L., & Mcloughlin, C. (1989). *Perspectives on teacher professional development*. London: Falmer Press.
Kitzinger, J. (1994). The methodology of focus groups: The importance of interactions between research participants. *Sociology of Health and Illness, 16*, 103–121.
Krueger, R. A., & Casey, M. A. (2000). *Focus groups: A practical guide for applied research* (3rd ed.). Thousand Oaks: Sage Publications.
Lai, K. (Ed.). (2001). *E-learning: Teaching and professional development with the internet*. Dunedin: University of Otago Press.
Lewis, K. (2002). *An enterprising future: Evaluating the young enterprise scheme*. Enterprise New Zealand Trust. Retrieved from http://sme-centre.massey.ac.nz/files/Complete_final_ENZT_report.pdf
Ministry of Education. (2007). *The New Zealand curriculum*. Wellington: Learning Media.
Ministry of Education. (2009). *Te Kete Ipurangi. What is education for enterprise?* Retrieved from http://education-for-enterprise.tki.org.nz/About-Education for Enterprise/Why-focus-on-Education for Enterprise/Defining-Education for Enterprise
Mitchell, L., & Cubey, P. (2003). *Characteristics of effective professional development linked to enhanced pedagogy and children's learning in early childhood settings: Best evidence synthesis*. Wellington: Ministry of Education.
Muller, J. (2011). Through other's eyes: The fate of the disciplines. In F. Christie & K. Maton (Eds.), *Disciplinarity: Functional linguistics and sociological perspectives* (pp. 13–34). London: Continuum.
Resnick, L. B. (1989, February). Developing mathematical knowledge. *American Psychologist, 44*(2),162–169. http://dx.doi.org/10.1037/0003-066X.44.2.162
Rhea, D. (2005). Bringing clarity to the "Fuzzy front end". A predictable process for innovation. In *Design research: Methods and perspectives* (pp. 145–154). Cambridge, MA: MIT Press.
Schön, D. (1991). *The reflective practitioner: How professionals think in action*. Aldershot: Ashgate.
Stoll, L., Bolam, R., McMahon, A., Wallace, M., & Thomas, S. (2006). Professional learning communities: A review of the literature. *Journal of Educational Change, 7*(4), 221–258.
Torp, L., & Sage, S. (2002). *Problems as possibilities: Problem-based learning for k-16 Education* (2nd ed.). Alexandria: Association for Supervision and Curriculum Development.
Ward, J. D., & Lee, C. L. (2002). A review of problem-based learning. *Journal of Family and Consumer Sciences Education., 20*(1), 16–26.
Waters, J. T., Marzano, R. J., & McNulty, B. A. (2003). *Balanced leadership: What 30 years of research tells us about the effect of leadership on student achievement*. Aurora: Mid-continent Research for Education and Learning.
Wenger, E. (1998). Communities of practice: Learning as a social system. *The Systems Thinker, 9*(5), 1–5.
Wood, D. (2007). Professional learning communities: Teachers, knowledge, and knowing. *Theory Into Practice, 46*(4), 281–290.
Young, M. F. D. (1998). *The curriculum of the future*. London: Falmer Press.

Chapter 13
A Synoptic View

David Barlex and P John Williams

Abstract This chapter provides an overview of the research findings reported in previous chapters, organising them into the following four categories: broad approaches to engaging pupils with designerly thinking; focused teaching and learning in technology education; classroom talk and communities of practice. For each of these categories the chapter then makes suggestions for further work and the contributions that teachers might make.

So, what are we to make of the findings embedded in each of the theses considered in this book? We organised the chapters into four groupings that we thought were of particular relevance to those who teach design and technology in schools:

- Broad approaches to engaging pupils with designerly thinking
- Focused teaching and learning in technology education
- Classroom talk
- Communities of practice

First this chapter will consider the findings of each grouping with a focus on their implications for classroom practice. Then this chapter will discuss the further work that might stem from the initial research and how teachers might contribute to this.

Findings Concerning Designerly Thinking

The narrative in England almost since the introduction of design and technology into the national curriculum in 1989 has been that the teaching of designing has been much less successful than the teaching of making. Over the first 10 years, the

D. Barlex (✉)
University of Exeter Associate, Brighton, UK
e-mail: D.Barlex@exeter.ac.uk

PJ. Williams
Curtin University, Perth, Australia
e-mail: pjohn.williams@curtin.edu.au

PJ. Williams, D. Barlex (eds.), *Contemporary Research in Technology Education*, Contemporary Issues in Technology Education, DOI 10.1007/978-981-10-2819-9_13

Office for Standards in Education consistently reported that skills in designing lagged behind those in making (Office for Standards in Education 1998, 2000). Recently Diana Choulerton, the National Lead for Design and Technology in England and a keynote lecturer for Design and Technology Association Summer School in July 2015, indicated that little had changed, reiterating the lack of success in teaching designing:

At Key Stage 3 (pupils aged 11–14 years) the curriculum consists of heavily guided making tasks with very limited opportunities to design in 3D. Very few opportunities to engage in an iterative design process.(Slide 7)

So after 25 years of teaching design and technology in England, the teaching of designing still leaves much to be desired. Informal conversations with colleagues in Europe, Canada, India, the USA, Australia and New Zealand indicate that they are experiencing a similar situation in their countries.

Hence research considering designerly thinking and how it might be improved is to be welcomed. The work of Farhat Ara probed Indian elementary and middle school students' ideas and images of design and designers and professional design activity revealing a limited, limiting and stereotypical view. In response she developed a theoretical framework for developing design activities and from this a set of relevant pedagogical principles. There is much in Farhat's work that could be applied to the teaching of designing anywhere in the world. Although dealing with older students, Matthew Watkin's research is also highly relevant. A key feature of designerly thinking is the need for critical reflection which Matthew associated with 'deep learning'. He devised interactive audiovisual workshops that provoked students to become more engaged and encouraged discussion. This led to increased critical reflection, a noted weakness of Net Generation learners. For those who teach design and technology in schools, this 'lack of critical reflection' will come as no surprise. Matthew's research indicating the immediate benefits of using the A/V resources and the styles of teaching he adopted are easily applicable to secondary learners. Jae Young's careful and thorough investigation of the problem-solving pupils needed to carry out during designing gives insight into the nature of support they need to be successful. The role of the teacher in providing this support is crucial, and unless the teacher herself understands the nature of designing, it is unlikely she will be able to provide this. Hence, Jae Young's research points clearly to the need for teachers themselves to have both design experience and competence. Many teachers will not have designed anything since their initial training so there is a strong argument for such experiences in ongoing professional development.

Findings Concerning Focused Teaching and Learning

In many technology curricula, there is an emphasis on developing contemporary technological content. It is difficult for schools to keep up to date given the rapid rate of technological change which shows no sign of abating. In England, for example, there has been a move away from designing and a concentration on the modernisation of content, particularly technical content. The Office for Standards in Education report for design and technology covering the years 2007–2010 commented that *too little use was made of electronics, computer aided design and manufacture (CAD and CAM) and control technology in the teaching of D&T* (p 5) and *It is critical that schools stay up to date* (p 47). To some extent, this criticism was endorsed by the expert panel set up to review the national curriculum (DfE 2011) which commented that design and technology had weak epistemological roots (footnote 58, p. 24). Most recently, Diana Choulerton in her keynote address for Design and Technology Association Summer School in July 2015 noted that *electronics and robotics were nowhere to be seen* (slide 7).

So there is a growing twofold concern that (a) there should be some technical content in what is actually to be learned and (b) that this content should be 'up to date'. It is clear that Deborah Winn's research addresses these concerns in developing and to some extent validating a new approach to teaching CAD that teachers with little previous experience can accommodate. Sarah Pule's research focused on electronics, a specifically mentioned area of weakness and concern. Her research endorsing the importance of an experiential approach to developing understanding can be used to inform the learning approaches in the early stages of electronics teaching within design and technology. The understanding of the behaviour of individual components may not be as significant as it once was with the increase in the use of programmable components to embed functionality into products in order to enhance and customise their operation. However such products will always contain components that are driven and activated by the embedded software. Understanding the behaviour of such components will be important, and Sarah's findings will be useful. Sammy Khosa explored the teaching of sectional drawing. Some might argue that this has been replaced by the use of CAD, but this would be to ignore the important understanding that lies within sectional drawing. Plans for making need to be interpreted whether these are generated by hand or machine, and if this interpretation is found wanting, then the making will be unsuccessful. So whilst we might want to extend the teaching of sectional drawing to include a CAD approach, it would be unwise not to use his finding that the use of real and physical examples enables learning to take place more effectively.

Findings Concerning Classroom Talk

The importance of classroom talk has long been recognised. The early and seminal writing of Douglas Barnes (1976) sets the stage for an appreciation of its significance. The constructivist approach to science education has its roots in provoking young people to talk and listening to what they say (Driver 1983; Sutton 1992). More recently Mercer and Hodgkinson (2008) produced an edited collection on ways of improving classroom talk which included pieces on mathematics and science. The meta-analyses of John Hattie (2009) indicate that classroom talk is one of the most effective ways of improving teaching. Yet there has been little research in classroom talk with regard to design and technology. Perhaps this is because many see the subject about 'doing' and time spent talking about 'what to do, how to do it and why' is time wasted. Nothing could be further from the truth. Working with very young children, Wendy Fox Turnbull's studies revealed that talking about relevant home experiences enabled the children to deploy knowledge that they had learned outside school, so bringing an extra richness to the design and technology classroom which in addition to enhancing the children's design and technology learning also gave teachers the opportunity to develop a deeper understanding of their pupils. Working with much older students, Deborah Stevenson's research probed the way students were able to use a sophisticated philosophical discourse to explore bioethical issues resulting in significant learning. The role of the teacher in introducing the necessary philosophical tools was essential, but equally important was the teacher's role in letting the students talk and develop understanding and insight for themselves. At a time when various new and emerging technologies are beginning to have serious consequences, it is important that design and technology teachers take the implications of Deborah's work seriously and give time to classroom talk.

Findings Concerning Communities of Practice

Being part of a community of practice when one is a teacher is inevitable but the extent to which that community sustains and enhances your practice can be highly variable. David Hargreaves (1998) notes the importance of a community of practice that supports the development of new ideas:

> *Good ideas – especially when they come from new or more junior colleagues – are fragile and may well need protection by the most experienced teachers from colleagues inclined to intellectual infanticide. Cynics kill knowledge creation.* (p. 31)

Some communities of practice can be highly transient although not necessarily ineffective because of that. The conversations and discussions taking place at conferences and professional development sessions between teachers who might never meet again can be the stimulus for curriculum development and new approaches to teaching. A more formal way of devising such transient communities is knot working (Engeström et al. 2012) in which a group is formed in response to a particular problem and work towards developing a solution. Once a solution has been developed and successfully implemented, the group (or knot) is dissolved. In this way, an organisation can respond fluidly and rapidly to problems as they emerge. Individuals are invited to become part of a knot because they have appropriate expertise or experience not because they have status or position in the organisation. Being part of a knot is therefore available to junior as well as senior members of the profession and can be a highly influential and positive experience for those new to the profession. This brings us to the work of Denise McGregor which studied the transition from a training environment at university to school. The role of mentoring was key to a successful transition and a successful start to a career in teaching. This is perhaps particularly important where training is limited to a single year and becoming single school based. The teaching experience will of necessity be limited, and the expectations of teachers and teaching in a first post might be significantly different to that in the training school. Here the role of a mentor in supporting induction to the community of practice will be crucial. Tome Maptose's research could not take place in a more different context than Denise's. Finding ways to support teachers who lack subject knowledge and are dealing with classes where the teacher–student ratio varies from 1:60 and 1:90 is without doubt a significant challenge. Yet Tome was able to show how such teachers could be supported by developing the Participatory Emancipation for Action Research (PEAR) model for use in those communities of practice. A key feature of the implementation of the PEAR model is collaboration between teachers from two schools thus widening the community of practice and as this community demonstrates success inviting further schools to join the activity and the community. This model of schools working together in clusters to raise student achievement is being applied in many jurisdictions. The work of Gary O'Sullivan investigates the interaction between two different communities: those working in technology education in schools and community groups in order to provide Education for Enterprise. Here we have an interesting example of a situation in which it is important to coalesce two different communities into one community of practice. Gary's research indicated the importance of student activities being based on local needs if the students were to have ownership of the resulting projects. Clearly the input of community groups can help in identifying appropriate activities, but unless they are part of the overall community of practice, it is difficult if not impossible for them to make this contribution.

Suggestions for Further Work and Teacher Contribution

Now we ask what further work might stem from the initial research and how teachers might contribute to this?

Suggestions Concerning Designerly Thinking

Farhat has two suggestions: (a) revealing views of designers held by young people in different cultures and the extent to which these contribute to their career aspirations and (b) what happens when the focus of the design activity shifts from the tangible to the intangible? The individual teacher can address the first part of (a) by asking their pupils to draw and annotate pictures showing a designer at work. These would provide immediate insight to inform the teacher's own practice, and a comparison across different cultures would indeed be fascinating. This is something that collaboration between design and technology teaching professional associations might achieve. Exploring the second part of (a) would require a longitudinal study, and in reality this can only be achieved by well-financed university-based research such as the ASPIRES Project (ASPIRES 2014) which looked at pupil aspirations in the light of their science education. Involvement in such a project would of course be professionally very rewarding for the teachers who took part. Changing the focus of design activity to embrace the design of services and experiences is an intriguing possibility as it involves moving away from using the skills of drawing, manipulation and actual making to requiring the skills observation, storytelling and acquaintance with new technology and media. The extent to which a teacher is able to engage pupils in such design activity will depend on the requirements of the programme of study in which they are engaged, but even a limited involvement is likely to be very informative. It might be couched in terms of critique and pupils identifying what they want to change about society and why. Reporting such activity as a case study in design and technology teaching professional association publications would be a first step which could be followed up by more extensive research, involving several case studies each from a different teacher perhaps, which could be presented at academic conferences.

Matthew's main concern is that his approach to developing deep learning might not be effective with younger students and it will be important to investigate this ideally through a longitudinal study which should probe (a) the extent to which his approach is effective and might need to be modified and (b) if modification is required what sort of guidance might be needed to achieve this. It is a relatively straightforward matter for a secondary school teacher to adopt Matthew's approach and find out the extent to which it is effective and deals with the 'Net Generation learning problem'. Professional association journals would be keen to report such work and possibly to publish associated resources which might incorporate additional or modified guidance. As indicated above, a longitudinal study would require much more resource.

Jae Young is concerned that teachers spend time developing their own insight into their pupils' design-based problem solving as a precursor to being involved in any further investigation. The intuitive understanding gained is important, as it will enable teachers to consider any related research in the light of their own experience. The focus of research that might take place might deal with the limitations of his own study: design-based problem-solving in the normal timetable and differences in this activity between various student groups. If a teacher were to write reflections on their pupils' design-based problem-solving over a period of time, the results could almost certainly be distilled into an article that would be published by a professional association journal, as it would be of great interest to many teachers. Such research activity might well find its place in an MA programme as the basis for a dissertation which could lead to publication in a peer-reviewed journal.

Suggestions Concerning Focused Teaching and Learning

Deborah noted that she was surprised by the way students with low academic ability were able to use the 'collaborative game approach' to become proficient at using CAD and producing highly creative outcomes, whereas academically able students sometimes seemed unable or unwilling to do so. So she wonders if it might be useful for some research to explore this apparent dichotomy and to develop approaches to effective paring. She felt intuitively that there might be some relationship between a pupil's existing spatial ability and their ability to use 3D CAD programmes and so believes it would be worth exploring this with a view to finding out the relationship between visualisation in the mind's eye and image manipulation on screen. Deborah gives strong support to the idea of teachers carrying out research through observing, reflecting on and evaluating their own practice. Hattie (2009) has reported that such formative evaluation is extremely effective in improving pupils' learning. Hence Deborah's suggestion that a teacher might try different strategies and evaluate their impact is well worth considering. Professional associations would be pleased to publish such research. She also advocates in-depth case studies to explore the way some pupils are particularly successful. Such work would probably require collaboration with a university department of education but could lead to peer-reviewed publications. And she wonders about the development of a spatial ability test which might be used to reveal the different starting points of particular students. This is a very demanding activity requiring expert input but one in which interested teachers could participate.

Sarah's research has led her to see learning activities that embrace 'embodied cognition' as being particularly significant for design and technology both in the general sense that designing and making rely heavily on 'doing' and the particular situations where the learning of specific technical concepts can be enhanced by hands-on experiences especially those devised to engage the learner with the concepts. Clearly a teacher can identify topics that pupils are finding difficult to comprehend and develop 'items to handle and explore' to help overcome the learning difficulties. Other teachers would welcome the reporting of such work, and it would easily find publication in professional association journals. An interesting feature of such work is how pupils reveal their enhanced understanding or lack of it. Both comprehension and confusion are likely to be revealed by gesture, expression and body language – in line with the way the pupils are or are not learning. Being aware of these signals is important for teachers who have to make rapid judgements about who is and is not being successful in a busy classroom. Capturing moments of comprehension and confusion on video in both informal (e.g. as they happen spontaneously) and formal (e.g. response to the teacher's questions) situations would provide a highly useful resource. Of course it would be necessary to gain permission to acquire and use such video from the pupils and their parents.

Sammy's research revealed the importance of developing spatial ability if students were going to make progress in technical drawing. Spatial ability of course has applications in fields other than technical drawing, and Sammy's research has resonance with both that of Deborah and Sarah. The use of 3D models to support technical drawing relates to Sarah's work on providing concrete models representing phenomenon, and Deborah's concern with spatial ability mirrors Sammy's view that diagnosing students' spatial ability is a key to technical drawing. Sammy argues that becoming proficient in technical drawing will develop students' spatial ability, an accomplishment useful in a wide range of subjects including the STEM subjects.

Hence developing alternative schemes of work and evaluating their effectiveness in developing spatial ability through learning in different subjects would be a useful research and development exercise that could be carried out by individual teachers. Collecting such work into a coherent form and publishing in one or more professional association journals would be a useful exercise that could be carried out by a group of teachers in a single school.

Suggestions Concerning Classroom Talk

Wendy raises a series of interesting and relevant questions that might be addressed in response to her research, but underlying these is the role of the teacher in technology lessons in which students are designing and making. From this I think is the question of timing and intervention, so it is neither too soon and hence compromises the pupil's ownership of the endeavour nor too late by which time the student has become disengaged through frustration. It would be useful to know what clues the teacher might note that indicate the 'right time to strike'. Facial expression, body language, moving away from the work piece and off-task talk can all indicate a pupil's state of mind with regard to the task they are tackling. Asking a question in response to such clues and listening carefully to the answer might be a more useful approach than simply interpreting the visual clues and intervening with advice. There is however very little research into this aspect of teaching design and technology, so it would be good to capture examples of 'state of mind' clues as both still images and video clips. This data in itself is unlikely to reveal much unless it can be situated in the narrative of the student's designing and making activity. It is the teacher that has access to the narrative and is able to interpret the data. So it would be extremely useful for a teacher to organise the capturing of the clues (through still photography or video filming of a student working) and annotating the results with a narrative of the activity. Of course it would be necessary to gain permission to acquire and use still or video images from the pupils and their parents. This would provide the basis for highly informative case studies dealing with the vexed question of 'knowing when to intervene'. There is little doubt that a professional association journal would publish such case studies and that they would find considerable use in initial teacher training.

Deborah raises the interesting question of assessing students' responses to value positions suggesting that they should not be judged by comparison with a given and perhaps preferred value position of the assessor but whether the position is well supported or not by the arguments provided by the students. Exactly how to do this is still an uncertain territory so well worth exploring. Deborah was also concerned that teachers felt inhibited with regard to using practical, student-centred and narrative- and discussion-based pedagogy in their classrooms and that this was worthy of further study, particularly a study that might enable teachers to develop experience of and expertise in such pedagogy. These two issues are related in that research carried out by teachers to develop their pedagogy could ultimately be collated and used to inform the development of assessment practice with regard to students' ability to support a particular value position. The individual research into pedagogy would be welcomed by professional association journals, and once a collection of case studies had been published, an awarding organisation would be able to use them as a springboard for devising appropriate assessment.

Suggestions Concerning Communities of Practice

Denise's research has indicated the importance of a sound mentoring programme during the first year of a teacher's career. However it is clear that the provision of such mentoring is highly variable so finding out what mentoring programmes exist in schools and the extent to which they are useful is an obvious area for further research. It is also worth noting that the initial training received by teachers can vary widely in length so the nature of mentoring programmes suited to different pre-service experiences would be a useful area of investigation as well. Schools wishing to gain the best from new entrants to the profession and 'set them fair' on a teaching career would do well to scrutinise their mentoring and its effectiveness. Teachers who are mentors and mentees have a significant role in this scrutiny. Professional associations for design and technology teachers have more than a vested interest in ensuring that new entrants to the profession are well cared for and do not become 'poor retention' statistics. Hence it would make sense for such professional associations to instigate research into this area and involve experienced teachers of the subject in taking an active part in mentoring.

Tome makes the important point that through action research teachers can take control of the research and investigation of their own teaching and in doing this they can certainly contribute to an understanding of practice and how it might be improved. This was particularly important in Tome's situation in which physical resources were scarce and classes large, but there is a general case to be made for the collaborative use of action research as an intervention strategy. In Tome's situation, it was necessary for someone to act as a convenor to initiate and oversee the intervention, but in the main the teachers themselves were in charge of the research and its dissemination. To my mind, this is a model that can easily be transferred to other situations. I can envisage teachers working in different schools in an academy chain in England, in a school district in the USA and in a local authority in Sweden collaborating in this way and developing enhanced practice. A key figure is the initiator, but I am sure that from within the ranks of teachers wishing to improve their practice, such individuals exist. In situations where there are professional associations for design and technology teachers, then it would be easy for the associations to instigate such activity and publish the findings in their journals.

Gary has noted that Education for Enterprise activities that take place as school–community partnerships are very context dependent and vulnerable to local issues. Although his research teased out factors that contributed to the success of the partnership, it must be acknowledged that these come from a small study. This does not in any sense invalidate them, but one must err on the side of caution in suggesting that such factors will necessarily play out in a similar fashion in other contexts. Hence it would make sense to gather together the findings from similar research in a range of different contexts and carry out a meta-study to discern the extent to which the various factors might be isolated from the impact of context. Meta-studies are very much the concern of the professional academic researcher concerned, so where does that leave the contribution of teachers? Two possibilities spring to mind. First there are opportunities for teachers to participate in Education for Enterprise school–community-based partnerships and report on their experience. If this activity is undertaken in collaboration with university-based researchers, then it will be of a quality that ensures it can contribute to the studies that inform a meta-study. Second those teachers who take part in Education for Enterprise school–community-based partnerships can use the results of the meta-study to guide their endeavours so that their work is not as context vulnerable as it might be otherwise.

End Note

To summarise, the PhD research carried out by the contributors to this book is clearly of value, and teachers can certainly contribute to taking the research further. The teacher professional associations are in a position to support teachers in this endeavour through their own publications and conferences. University departments of education can support teachers on their MA programmes to carry out and report research activities that make further contributions to the fields explored by the PhD studies.

References

ASPIRES. (2014). http://www.kcl.ac.uk/sspp/departments/education/research/aspires/ASPIRES publications.aspx

Barnes, D. (1976). *From communication to curriculum*. Harmondsworth: Penguin.

Choulerton, D. (2015). Design & Technology Association Summer School keynote address can be found at this url: http://www.slideshare.net/Ofstednews/design-and-technology-association-data-summer-school-keynote-2015

Department for Education. (2011). *The framework for the national curriculum. A report by the expert panel for the national curriculum review*. London: Department for Education.

Driver, R. (1983). *The pupil as scientist*. New York: Taylor and Francis.

Engeström, Y., Kaatrakoski, H., Kaiponen, P., Lahikainen, J., Laitinen, A., Myllys, H., et al. (2012). Knotworking in academic libraries: Two case studies from the University of Helsinki. *Liber Quarterly, 21*(4), 387–405.

Hargreaves, D. (1998). *Creative professionalism. The role of teachers in the knowledge society*. London: Demos.

Hattie, J. (2009). *Visible learning*. Oxford: Routledge.
Mercer, N., & Hodgkinson, S. (2008). *Exploring talk in schools*. Thousand Oaks: Sage.
Office for Standards in Education. (1998). *Secondary education-A review of secondary schools in England 1993–1997*. London: The Stationary Office.
Office for Standards in Education. (2000). *Ofsted subject reports secondary design and technology, 1999–2000*. London: The Stationary Office.
Office for Standards in Education. (2011). *Meeting technological challenges design and technology in schools 2007–10*. London: The Stationary Office.
Sutton, C. (1992). *Words, science and learning*. Buckingham: Open University Press.

Lightning Source UK Ltd.
Milton Keynes UK
UKOW05n2337231216
290788UK00011B/85/P